EARTH SCIENCE

A Scientific History of the Solid Earth

DISCOVERING the EARTH

EARTH SCIENCE

A Scientific History of the Solid Earth

Michael Allaby

Illustrations by Richard Garratt

Facts On File
An imprint of Infobase Publishing

EARTH SCIENCE: A Scientific History of the Solid Earth

Facts On File, Inc.
An imprint of Infobase Publishing
132 West 31st Street
New York NY 10001

Library of Congress Cataloging-in-Publication Data
Allaby, Michael.
    Earth science: a scientific history of the solid Earth / Michael Allaby; illustrations by Richard Garratt.
        p. cm.—(Discovering the earth)
    Includes bibliographical references and index.
    ISBN-13: 978-0-8160-6097-9
    ISBN-10: 0-8160-6097-5
    1. Earth sciences. I. Title.
    QE26.3.A45 2009
    550—dc22   2008016780

Facts On File books are available at special discounts when purchased in bulk quantities for businesses, associations, institutions, or sales promotions. Please call our Special Sales Department in New York at (212) 967-8800 or (800) 322-8755.

You can find Facts On File on the World Wide Web at http://www.factsonfile.com

Text design by Annie O'Donnell
Illustrations by Richard Garratt
Photo research by Tobi Zausner, Ph.D.

Printed in China

CP FOF 10 9 8 7 6 5 4 3 2 1

This book is printed on acid-free paper.

CONTENTS

CHAPTER 8
DRIFTING CONTINENTS AND PLATE TECTONICS 178

PREFACE

Almost every day there are new stories about threats to the natural environment or actual damage to it, or about measures that have been taken to protect it. The news is not always bad. Areas of land are set aside for wildlife. New forests are planted. Steps are taken to reduce the pollution of air and water.

Behind all of these news stories are the scientists working to understand more about the natural world and through that understanding to protect it from avoidable harm. The scientists include botanists, zoologists, ecologists, geologists, volcanologists, seismologists, geomorphologists, meteorologists, climatologists, oceanographers, and many more. In their different ways all of them are environmental scientists.

The work of environmental scientists informs policy as well as providing news stories. There are bodies of local, national, and international legislation aimed at protecting the environment and agencies charged with developing and implementing that legislation. Environmental laws and regulations cover every activity that might affect the environment. Consequently every company and every citizen needs to be aware of those rules that affect them.

There are very many books about the environment, environmental protection, and environmental science. Discovering the Earth is different—it is a multivolume set for high school students that tells the stories of how scientists arrived at their present level of understanding. In doing so, this set provides a background, a historical context, to the news reports. Inevitably the stories that the books tell are incomplete. It would be impossible to trace all of the events in the history of each branch of the environmental sciences and recount the lives of all the individual scientists who contributed to them. Instead the books provide a series of snapshots in the form of brief accounts of particular discoveries and of the people who made them. These stories explain the problem that had to be solved, the way it was approached, and, in some cases, the dead ends into which scientists were drawn.

There are seven books in the set that deal with the following topics:

- Earth sciences,
- atmosphere,
- oceans,
- ecology,
- animals,
- plants, and
- exploration.

These topics will be of interest to students of environmental studies, ecology, biology, geography, and geology. Students of the humanities may also enjoy them for the light they shed on the way the scientific aspect of Western culture has developed. The language is not technical, and the text demands no mathematical knowledge. Sidebars are used where necessary to explain a particular concept without interrupting the story. The books are suitable for all high school ages and above, and for people of all ages, students or not, who are interested in how scientists acquired their knowledge of the world about us—how they discovered the Earth.

Research scientists explore the unknown, so their work is like a voyage of discovery, an adventure with an uncertain outcome. The curiosity that drives scientists, the yearning for answers, for explanations of the world about us, is part of what we are. It is what makes us human.

This set will enrich the studies of the high school students for whom the books have been written. The Discovering the Earth series will help science students understand where and when ideas originate in ways that will add depth to their work, and for humanities students it will illuminate certain corners of history and culture they might otherwise overlook. These are worthy objectives, and the books have yet another: They aim to tell entertaining stories about real people and events.

—Michael Allaby
www.michaelallaby.com

ACKNOWLEDGMENTS

All of the diagrams and maps in the Discovering the Earth set were drawn by my colleague and friend Richard Garratt. As always, Richard has transformed my very rough sketches into finished artwork of the highest quality, and I am very grateful to him.

When I first planned these books I prepared for each of them a "shopping list" of photographs I thought would illustrate them. Those lists were passed to another colleague and friend, Tobi Zausner, who found exactly the pictures I felt the books needed. Her hard work, enthusiasm, and understanding of what I was trying to do have enlivened and greatly improved all of the books. Again I am deeply grateful.

Finally, I wish to thank my friends at Facts On File, who have read my text carefully and helped me improve it. I am especially grateful for the patience, good humor, and encouragement of my editor, Frank K. Darmstadt, who unfailingly conceals his exasperation when I am late, laughs at my jokes, and barely flinches when I announce I'm off on vacation. At the very start Frank agreed this set of books would be useful. Without him they would not exist at all.

INTRODUCTION

Not far from the village of Brandon, in the county of Norfolk in eastern England, there is an area of about 96 acres (39 hectares [ha]) of uneven, grass-covered land with pits, abandoned quarries, spoil heaps, and more than 400 deep holes. The Anglo-Saxons, who colonized England after the departure of the occupying Romans, called the place Grim's Graves, after their god Grim. It was also known as the Devil's Holes. Today it is called Grime's Graves, but it is not a graveyard and the holes are not graves.

BENEATH OUR FEET

It was not until 1870 that archaeologists began to study the area. They discovered that Grime's Graves is a 5,000-year-old industrial site. The holes are mine shafts, dug by Neolithic (New Stone Age) miners using picks made from the antlers of red deer. The miners were extracting jet-black flint, which they found about 30 feet (9 meters [m]) below ground level. Horizontal galleries radiating from the bottoms of the shafts follow the seams of flint. The area is an important archaeological site, managed by English Heritage and open to the public.

Flint was used to make cutting tools and weapons such as arrowheads. More recently it was used to make the sparks that fired flintlock muskets. It was a valuable resource, mined and worked at places like Grime's Graves and traded widely. Eventually it fell from use, replaced by metals that make better tools with sharper points and edges.

Grime's Graves provide clear evidence—if it were needed—of the extent to which people have always depended on the rocks beneath their feet. As well as tools and weapons, rocks provide stone and clay bricks for building, slate for roofing, and stone to build walls that enclose livestock and protect them from predators. Monuments and ceremonial buildings are constructed from large stones. Stonehenge in England is built from stones and so are the Egyptian pyramids and the Greek Parthenon. Metals are extracted from ore rocks. Bright gemstones that make jewelry and ornaments for the crowns of monarchs are *minerals,* cut and polished, but first found in rocks.

People are inventive. Someone, long ago, found that striking a piece of flint in a particular way produces a fragment with a sharp edge. People are also curious. We cannot know whether the miners at Grime's Graves speculated about the nature of their flint—wondered what it is made from and how it came to be embedded in the chalk rock—for they left no written record. It would be surprising if they did not speculate, however, because people's curiosity leads them to ask questions about the world they inhabit. They delight in stories and search for explanations for the objects they find and the phenomena they observe. So, the study of the Earth and its rocks is very ancient. The most familiar name for that study is *geology,* derived from two Greek words: *ge,* which is one version of *gaia* and means "Earth," and *logos,* meaning "word," "reason," or "account." Geology is an account of the Earth.

As the study of the Earth developed over the centuries, geology began to divide into separate disciplines. The aspects of most interest to physicists became geophysics, and geochemists specialized in studying the chemical reactions that take place below ground. Geomorphologists studied the development of landforms visible at the surface, mineralogists studied minerals, seismologists studied earthquakes, volcanologists studied volcanoes, petrologists studied rocks and their origins, and several more disciplines developed. All of these are now grouped together as the Earth sciences, also called geoscience or the geosciences. The Earth sciences concern one part of the natural environment, so they form part of the larger grouping of environmental sciences. This book is about the Earth sciences.

Some scientists use the term very broadly, regarding climatology, meteorology, and oceanography also as Earth sciences. In this book the term is used more restrictively to describe only the study of the solid Earth.

Earth Science begins with the aspect of the Earth of most interest and importance to travelers, explorers, adventurers, and merchants: How large is the Earth, and how are its lands and seas distributed? Chapter 1 tells of how the size of the planet came to be measured, and chapter 2 tells of the way its shape was determined and its surface mapped. Having determined the general appearance and dimensions of the surface, chapter 3 outlines early ideas about what lies beneath the surface. Is the Earth filled with water? Is it hollow? Why are there volcanoes and earthquakes?

From earliest times people have used metals. Long before the invention of metal tools, the wealthy and powerful possessed gold ornaments. Chapter 4 describes how people learned to extract metals from the Earth's ores. It also tells of the age-old link between precious metals and power, recounting the tales of the Golden Fleece and El Dorado.

Certain rocks contain fossils. These were long regarded as curiosities, but chapter 5 explains how their true nature was discovered and the implications of that discovery for the history of the Earth. The study of fossils led to the realization that Earth has a history that began a very long time ago. Chapter 6 recounts the steps by which the history of the Earth was teased from the rocks, and it explains the rival theories of catastrophism and uniformitarianism as well as neptunism and plutonism, all of which were advanced to account for the origin of the rocks found at the Earth's surface. The chapter ends by telling how the Earth's history came to be divided into the episodes making up the geologic time scale and includes the present version of that time scale.

Mountains are made from rocks that appear to have been folded, tilted on end, and crumpled, and many of those rocks contain the fossils of shellfish. Various hypotheses were proposed to explain the origin of mountains. The most enduring of these held that the Earth was once molten and that throughout its history it had been gradually cooling. As it cooled, the Earth contracted, and as it contracted, its crust shrank and crumpled like the skin of an old, dry apple. Many years passed before this idea was finally dispelled, only to give way to an idea that seemed still more preposterous: The Earth's continents move about and collide with each other. Chapter 7 explains the competing ideas about the way in which mountains form, and chapter 8 describes the development of the theory of plate tectonics, which explains mountain building and much else besides.

Plate tectonics is the unifying theory that binds all of the Earth sciences together. Appropriately, therefore, this is the last chapter. It marks the point that the story of the Earth sciences has now reached. Earth's story has not ended, nor has the research leading to an ever-deeper understanding of it, but the rest is yet to come.

This book has been great fun to write. I hope it is fun to read.

—Michael Allaby
Tighnabruaich, Scotland
www.michaelallaby.com

Measuring the Earth

Nowadays the captains of ships and pilots of airliners have orbiting satellites to monitor their positions. They navigate by GPS (global positioning system). Even car drivers, long-distance hikers, and mountain climbers use GPS.

Before GPS became available, people used maps and the stars. Sailors measured their latitude by the positions of stars. Long-range airplanes flying at night, including bombers in World War II, had a plastic "bubble" on the top of the fuselage from which the navigator had a clear view of the stars. The bubble was called an "astrodome," to reflect its purpose. Maps of Europe, the United States, and many other parts of the world were detailed and accurate.

Navigation is now so straightforward that it is easy to forget just how recent these developments are. It was not until the 18th and 19th centuries that astronomers and surveyors had the tools and knowledge to draw accurate and detailed maps of parts of the United States, England, France, and India. This chapter explores the first of the difficulties mapmakers had to overcome. Before they could draw their maps they had to determine the shape and size of the Earth. There are many places the story might start, but one of the most famous of all maritime adventures and navigational disasters is as good as any. Let the story begin with Christopher Columbus.

HOW CHRISTOPHER COLUMBUS DID NOT FIND JAPAN

Half an hour before sunrise on August 3, 1492, three small ships sailed out of the port of Palos, on the coast of the Gulf of Cádiz in southern Spain, not far from the modern city of Huelva. The party reached the Canary Islands on August 12 and departed from there on September 6, heading out into the broad Atlantic Ocean. The three vessels were the *Pinta,* commanded by Martín Alonso Pinzón, the *Niña,* commanded by his brother, Vicente Yáñez Pinzón; and the *Santa María,* commanded by a very experienced Genoese-born sailor, Cristoforo Colombo (Hispanicized to Cristóbal Colón), known to the English-speaking world as Christopher Columbus (1451–1506), who was the leader of the expedition. He may have belonged to a Spanish-Jewish family living in Genoa and he wrote only in Spanish or Latin.

Columbus's aim was to reach Asia by traveling westward rather than eastward. Asia was the source of many valuable commodities, especially gold and spices, but the journey to these fabulous riches was long and hazardous. Ships sailing from Europe had to travel around the continent of Africa and through the storms of the Cape of Good Hope before braving the typhoons of the Indian Ocean. Rather than take the risk, Europeans imported Asian goods along an overland trade route that consisted of a chain of merchants. This system worked well enough for many years, but during the 15th century the Ottoman Turks, who until then had ruled only northern Turkey, expanded their empire, encompassing the trade routes. The Turks imposed heavy duty on goods passing through their territory, and the trade between Asia and Europe declined as the cost of imports rose. Clearly, rich rewards awaited any European merchant or sea captain who could find a way to bypass the Turks.

The idea of sailing westward to Asia was not entirely original. Several other would-be explorers had discussed it before Columbus developed it into a practical scheme and persuaded Ferdinand and Isabella, the king and queen of Spain, of its value.

A deeply pious man, Columbus found justification for his plans in various scriptural passages that he interpreted as predictions of success. His extensive reading of the accounts of travelers, as well as of the Bible, led him to conclude that the Earth is spherical, the surface of the Earth is covered by six parts dry land and one part ocean, and the distance between Spain (the edge of the West) and India (the edge of the East) is very long by land but very short by sea. Columbus reckoned that traveling eastward by land across Europe and Asia, the distance between Spain and India was 282° of longitude. There are 360° of longitude in all, so Columbus surmised that the distance between Spain and India traveling westward by sea must be 78° (360° – 282°). That being so, Columbus calculated the distance to be 2,760 miles (4,440 kilometers [km]).

Columbus had a map to help him prepare. The original version had been drawn by Ptolemy (Claudius Ptolemaeus), an astronomer and geographer, probably Egyptian, who lived in Alexandria in the second century C.E. It had appeared in Ptolemy's book *Geographia*, but Italian cartographers had subsequently greatly modified it. The map suggested the possibility of reaching India by sailing westward. Columbus also had a chart to help him navigate prepared by Paolo Toscanelli (1397–1482), a Florentine physician and mapmaker. Toscanelli based his chart on Ptolemy's map, embellished it with travelers' tales and legends, and showed the Atlantic Ocean with Europe in the east and Asia in the west.

As the days dragged on and the three ships continued westward, Columbus realized they must have covered about 5,175 miles (8,327 km) rather than the 2,760 miles he had anticipated. He concluded that the Earth must be larger than was shown on his chart. Nevertheless, when his increasingly scared and mutinous crew finally espied land, two hours after midnight on October 12, Columbus had not the slightest doubt where they were. He named the first island they reached San Salvador, claimed it for Spain, and was convinced it was one of the outlying islands close to Cipango (Japan). He imagined the local people were subjects of a great king who lived on a large island they called Cuba, which he assumed was Cipango. The island he named San Salvador was Guanahani, in the Bahamas, and the people he met were definitely not Japanese.

Columbus was wrong on every count, but this was not his fault. He was a skilled and brave sailor who did the best he could with the knowledge and tools available to him. He lacked only two things: an accurate measure of the size and shape of the Earth and a reliable chart based on that measure. Navigators would have to wait many years for either of these.

IS THE EARTH A DISK OR A SPHERE?

Despite his errors and those forced on him by the false information available to him, Columbus was a keen observer and experienced navigator. Like all explorers, he charted the coasts of the lands he encountered, using the Pole Star to measure his latitude. Measuring longitude was much more difficult. Seen from anywhere in the Northern Hemisphere, the Pole Star is directly above the North Pole, so the direction toward it is always north. Measure the angle of the Pole Star above the horizon, and that angle is equal to the latitude of the observer. Navigators can also use the Sun and many other stars to calculate latitude by measuring the body's angle of elevation, the *declination,* at its highest point in the sky.

One night during his third voyage to the West Indies (1498–1500), Columbus was measuring the strait between Trinidad and Venezuela. He knew the distance between them was less than 70 miles (113 km), and he knew the length of a degree of latitude. But when he measured the latitudes he found that the Venezuelan coast was at almost 5°N and the coast of Trinidad was at almost 7°N. It was impossible for two places so close together to be separated by as much as two degrees of latitude unless the Earth was what Columbus described as "deformed." In other words, it was not a perfect sphere.

No one by that time supposed that the Earth was flat. The story that Columbus held a minority view in believing the planet to be spherical is quite wrong. It is true that astronomer-priests of many early civilizations had believed the world to be flat (see sidebar), and the ancient Greeks believed that the Earth was supported by four elephants standing on the back of a great turtle—though they never offered any suggestion about what the turtle rested on. As early as the sixth century B.C.E., however, at least some Greek philosophers accepted that the world is spherical. Pythagoras (ca. 580–ca. 500 B.C.E.), a religious philosopher and mathematician, may have been the first person to propose a spherical Earth. Aristotle (384–322 B.C.E.) and Hipparchus (ca. 190–ca. 120 B.C.E.) certainly accepted the idea.

The measurement Columbus made of the strait between Trinidad and Venezuela challenged the traditional view, not that the Earth is a sphere, but that it is a perfect sphere. The Greek philosophers taught that geometry determined the shapes and relationships of objects in the universe and that this cosmic geometry was perfect. The spherical

BELIEF IN A FLAT EARTH

During a lunar eclipse the shadow of the Earth crosses the Moon's disk. The shape of the Earth's shadow is circular. Astronomers who know what causes an eclipse should be able to see that the shadow is of a circular object, most probably a sphere, and during the eclipse the spherical shape of the Moon is clearly visible and unmistakable. When a ship approaches across the horizon or a distant traveler comes into view across a vast plain, the object appears to rise above the horizon. The top of the mast or the head of the traveler appears first. This fact, too, might suggest that the Earth is spherical and that the horizon is the limit beyond which the curved surface falls from view.

Astronomers who undertake long journeys northward or southward can hardly help noticing another phenomenon: Stars to the south appear lower in the sky the farther north the astronomer travels. Again the most plausible explanation is that an observer's line of sight to the horizon is a tangent to the surface of a sphere and that the angle by which a star is elevated above that line depends on the observer's location on the sphere.

In the diagram illustrating this, two observers at different points see the same star, but it appears much higher in the sky to one observer than it does to the other.

Despite this, both the Babylonians and the ancient Egyptians believed that the Earth is a flat disk. Both civilizations were fascinated by the stars, and their priests were keen students of astronomy. The earliest reference to the names of galaxies was written in about 1700 B.C.E. by a Babylonian priest, and cuneiform inscriptions on a series of three clay tablets called Mul.Apin refer to more than 30 constellations. Those tablets were inscribed in about 1100 B.C.E. by or under direction from astronomer-priests who believed the Earth to be flat. Homer, the Greek poet who lived some time between 900 B.C.E. and 800 B.C.E. and wrote the *Iliad* and *Odyssey*, believed that the world was a convex dish surrounded by a river called Oceanus. Some Greek philosophers thought that the world journeyed through the heavens supported by four elephants that stood on the back of a giant turtle.

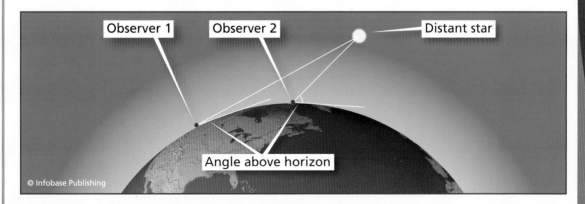

Using a distant star to show that the Earth is spherical. Two observers in different locations see the same star, but to one observer it appears higher above the horizon than it does to the other.

Earth was necessarily a perfect sphere. Aristotle shared this view and so did the Catholic Church: God made the world spherical, and God would not make the sphere less than perfect. Columbus had made a discovery with wide implications.

Columbus could have been mistaken. He used a quadrant to measure the angle of declination of the Pole Star, and although he had used the instrument many times before and found it reliable, perhaps he misread it slightly, or perhaps it had been damaged and was slightly out of alignment.

A quadrant is a simple instrument. As the diagram shows, it consists of a quarter circle—a quadrant—bearing a graduated scale calibrated in degrees, minutes, and seconds along the arc and with a movable arm pivoted at the center of the circle. A plumb line hangs from the center. The person using the instrument first makes sure that the plumb line hangs vertically, down the center of the vertical arm of the quadrant; the horizontal arm then points directly to the horizon (even if the horizon is obscured). Holding the quadrant very steady, the observer next moves the arm until it points at the star and reads off the angle of declination on the graduated scale. Unless the plumb line is absolutely vertical the quadrant will give a false reading. Columbus must have known this and would not have made so elementary a mistake.

Errors could also arise from two factors of which no one in Columbus's day was aware. The first is that the atmosphere refracts sunlight. When an observer watching a sunset sees the lower edge of the Sun begin to disappear below the horizon, the entire Sun is in fact already below the horizon. It remains visible

The quadrant. Having ensured that the plumb line hangs vertically, the user aligns the movable arm with a star and reads the angle of declination from the graduated scale.

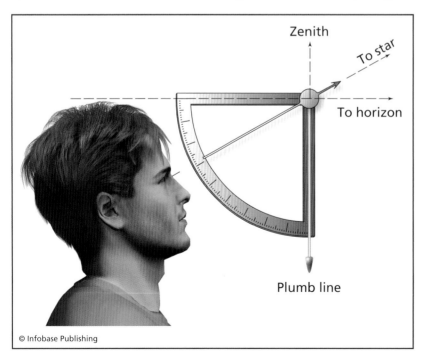

Zenith

To star

To horizon

Plumb line

© Infobase Publishing

because the atmosphere bends the light rays. The second source of error is due to the fact that mountains exert a gravitational force acting horizontally. The weight on a plumb line is deflected toward a mountain. The force is very weak and the deflection is tiny, but it is enough to make a sensitive instrument give a false reading. There are no large mountains between Trinidad and Venezuela, so this effect would not account for the discrepancy Columbus observed. Only one explanation therefore remains: As he reported, the Earth really is "deformed."

ERATOSTHENES—AND THE EARTH'S CIRCUMFERENCE

Several centuries would pass before the answer to Columbus's riddle of the "deformed" Earth was found. More immediately so far as Columbus was concerned, why was his first voyage across the Atlantic so much longer than he had anticipated? The answer to that is quite simple: The Earth is bigger than he had imagined. Columbus reached the same, rather obvious conclusion and made allowance for it in his subsequent voyages, but apart from revising his estimate of the time it took to sail across the Atlantic, he had no way of measuring the size of the entire Earth. It had been measured, centuries earlier. In fact, it had been measured twice, once almost correctly and once incorrectly. Unfortunately, Ptolemy used the incorrect value, which is why the map Columbus used greatly underestimated the width of the ocean.

The Greeks were the first people to attempt the task of measuring the Earth. Before they could set about making measurements, however, their thinkers had to accept and embrace a truly radical idea: The Earth is a physical entity, an object with shape and dimensions. That seems obvious today, and cameras on spacecraft have taken photographs showing the planet isolated—and clearly defined—in the vast blackness of space. It was not at all obvious until someone proposed the idea and produced reasons for believing it. People see the world around them. The world contains objects, such as rocks and trees, but no one could imagine being so far removed from it as to see the entire world as an object in itself. But until thinkers could accept that idea they could not possibly jump to the idea of measuring it. They could (and did) measure the distance between cities and between the islands of the Adriatic, Ionian, and Aegean Seas, but it was once inconceivable that these seas and places existed within a larger context that also had dimensions.

It was the mathematician-philosophers whose line of reasoning led to the attribution of dimensions to the world. To a person who stands in the middle of a vast, open plain or on a ship at sea and out of sight of land, the distance to the horizon appears to be the same in every direction, implying that the observer stands at the center of a circle. The Greeks believed the world was a circle, but as they developed the concept it occurred to them that a world made by the gods in the form of a circle must be a perfect circle, and a perfect circle is a circle that can be rotated without its shape being altered. Rotate a circle and it describes a sphere; consequently the circular world must in fact be spherical.

Actually measuring the sphere was a formidable task. It would be impossible to lay a rope or tape measure all the way around the planet, and even if someone thought of a way to do so, the result would be hopelessly inaccurate because the Earth's surface is very uneven and the tape would have to go up hill, down dale, and across high mountains. But Eratosthenes had a better idea.

Eratosthenes (ca. 276–ca. 196 B.C.E.) was an astronomer, mathematician, grammarian, literary critic, historian, and geographer. Indeed, he was one of the world's first geographers. In about 220 B.C.E. he made a map of the region extending from the British Isles to India and Sri Lanka and from north of the Caspian Sea to Ethiopia. It included the names of the peoples inhabiting some of the lands shown. His map of the whole of the known world was better than any of its predecessors. His interests were almost boundless, but no one can excel at everything and Eratosthenes earned the nickname "Beta." Beta (B, β) is the second letter in the Greek alphabet, and in ancient Greece it was also the symbol for the number 2. The nickname implied that Eratosthenes was second best at many of the things he attempted—but second in the whole world, which turns the nickname into a kind of compliment. He was born at Cyrene, near the modern city of Shahhat, on the coast of Libya. He studied grammar in Alexandria, Egypt, and philosophy in Athens, and in 240 B.C.E. he was appointed librarian of the library at Alexandria. This was the world's greatest library, and Eratosthenes remained there the rest of his life.

According to the traditional account, Eratosthenes knew that near Syene (modern Aswān), Egypt, there was a deep well where on Midsummer Day the Sun at noon shone directly onto the water. Syene was very close to the tropic of Cancer. He then measured the declination of the Sun at noon on Midsummer Day at Alexandria. He did not look directly at the Sun, of course, because that would simply dazzle him and in any case there is a very real risk that looking directly at the Sun will cause permanent damage to the eyes. Instead he measured the length of the shadow cast by an obelisk, the height of which he knew, perhaps using an accurate instrument called a *skiotheron,* or "cloud catcher," and used trigonometry to calculate that the Sun was 7.2° from the *zenith,* the point directly overhead.

Eratosthenes also knew the distance between Alexandria and Syene. Egyptian pacers (men who measured distances by pacing them) and camel drivers had measured it as 5,300 stadia. The stadion was a unit of linear measurement that was widely used in the ancient world, but it had different values in different places and times. No one knows the precise modern equivalent of the stadion Eratosthenes used, but historians believe it is equal to between 500 feet (152 m) and 607 feet (185 m). Syene is not directly south of Alexandria, so Eratosthenes corrected the distance to 5,000 stadia.

There are 360° in a full circle. Eratosthenes divided 360° by 7.2° and found that the distance between Alexandria and Syene is one-fiftieth of the circumference of the Earth (360 ÷ 7.2 = 50). Multiplying the 5,000 stadia distance by 50 gave him a value of 250,000 stadia for the circumference of the Earth. Depending on the correct value for the stadion, this is equal to between 23,674 miles (38,092 km) and 28,740 miles (46,243 km). The correct length for the Earth's circumference is 24,902 miles (40,074 km), so Eratosthenes was correct to within between –5 percent and +15 percent. It was an amazing achievement.

POSEIDONIUS—AND WHY COLUMBUS THOUGHT HE HAD REACHED JAPAN

Not everyone was happy with Eratosthenes' calculation. In those days the world known to the Greeks consisted of the lands bordering the eastern Mediterranean; the kingdom straddling what are now northern Libya and Egypt ruled by the Ptolemies, a dynasty of Macedonian kings; the Seleucid kingdom; and the countries of Parthia and Bactria farther east. As the map shows, their world was not large. It seemed large to the Greeks, however, and that was the problem. If the distance around the world was really as much as 250,000 stadia, it meant that the known world amounted to no more than about one-quarter of the whole world, and much of the known world was covered by sea. Eratosthenes' world as he calculated it was therefore believed to be altogether too big.

Eratosthenes lived to the age of 80, but he became blind and weak and finally, in 196 B.C.E., deliberately starved himself to death. In about 135 B.C.E., some 60 years later, Poseidonius was born in Apamea, Syria, in a time of political anarchy following the end of the Seleucid kingdom that until then had ruled the region. He became a Stoic philosopher, teacher, and what would nowadays be called a scientist.

Poseidonius traveled widely, through Italy, the eastern Adriatic along what is now the coast of Croatia, North Africa, Gaul (modern France), and westward to Spain, conducting scientific research. When finally he became a teacher, establishing his school on the Aegean island of Rhodes, he was already famous, and his school attracted pupils from the wealthiest and most influential families.

His travels were undertaken largely in connection with his study of the tides. He believed that ocean tides are caused by the pull of the Moon. He was not the first person to hold this view: The Greek explorer and geographer Pytheas, who flourished in about 325 B.C.E., had also suggested it. The Mediterranean contains such a small volume of water that it has no tides. If he was to study tides, therefore, Poseidonius needed to journey westward, all the way to the shores of the Atlantic.

Poseidonius was the first astronomer to take account of the refraction of light by the atmosphere, and he also studied the Sun. Various philosophers had attempted to measure the diameter of the Sun in terms of the Earth's diameter. Aristarchus (ca. 320–ca. 250 B.C.E.) thought the diameter of the Sun was 180 times that of the Earth, and Hipparchus thought it was 1,245 times. Poseidonius calculated it as 6,545 times. The true figure is that the Sun's diam-

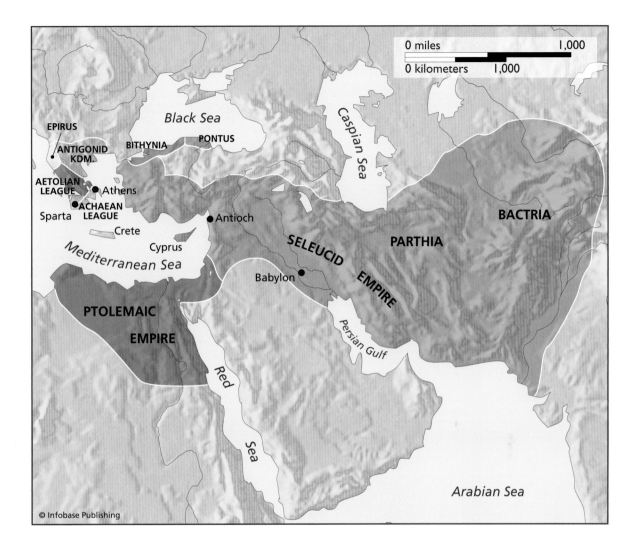

eter is 11,726 times that of the Earth. Poseidonius was wrong, but his estimate was much closer than that of anyone else up to that time.

Then Poseidonius "corrected" Eratosthenes' estimate for the Earth's circumference. He repeated the work Eratosthenes had done, but with some refinements and found that the circumference of the Earth is about 18,000 miles (28,960 km). This is the value Ptolemy accepted and on which he based his map. Ptolemy consequently depicted the Earth as being almost 30 percent

The Greek world in the third century B.C.E.

smaller than it is. No one has any idea why Ptolemy used this incorrect figure rather than Eratosthenes' estimate. What is even more surprising is that this value was still being used as late as 1492, 1,600 years later.

Mapping the Earth

Someone who sets off from home to visit a local store, perhaps followed by a visit to a hairdresser, and then to meet a friend for lunch in a favorite restaurant has no trouble finding the way. Traveling short journeys over familiar ground is simple, straightforward, and carries no risk of becoming lost. That is because everyone carries a mental record of his or her local surroundings.

People navigate locally by using landmarks. The person off on a shopping trip knows to turn left at the church and then right at the third traffic light. Prominent buildings, road junctions, and traffic lights are landmarks. In other parts of the world the landmarks may be harder to recognize, at least for someone unfamiliar with the landscape. Traditionally, Inuit families hunted for food across the frozen sea or inland through the tundra. In the Sahara nomadic Bedouin would traditionally spend the winter rainy season driving their livestock from place to place in search of good pasture. Mongolian nomads follow a regular routine that takes them from one grazing area to another. All of these people move through the landscape confidently and without hesitation. They know precisely where they are going, and they are familiar with every landmark.

Polynesian peoples also traveled long distances, in their case by sea, where they were often out of sight of land for days at a time. Their landmarks were in the sky and on the sea. The shape and direction of clouds, the position of the Sun, Moon, and stars, the appearance of the sea, all helped them navigate across the Pacific Ocean.

Most people are able to draw a representation of their journeys that would be good enough to guide someone wishing to follow the same route. A graphical representation of an area is a map. Almost anyone can draw a simple map. When people began to travel farther afield, however, the simple map was no longer sufficient. Travelers needed a map of a much larger area of the world, or even of the whole world—or at least the whole known world.

This chapter is about maps. It outlines their history and explains how they are made. Before anyone can draw a really reliable map, the curious anomaly Columbus discovered needs to be laid to rest.

OBLATE OR PROLATE?

Columbus discovered that Trinidad and the coast of Venezuela are separated by two degrees of latitude, even though he knew that the distance between them is less than 70 miles (113 km). He concluded that the Earth is "deformed." If that "deformation" is real, it is of great importance to anyone wishing to draw a map, because it suggests that one degree of latitude varies in length from one part of the world to another.

In 1687 Sir Isaac Newton (1642–1727), the English scientist and mathematician, published his greatest work, *Philosophiae naturalis principia mathematica* (Mathematical principles of natural philosophy), usually known simply as the *Principia.* The *Principia* is about motion and the force of gravity, and in it Newton stated that because the Earth rotates about its axis, almost certainly its shape is deformed: Its diameter is longer at the equator than it is from pole to pole. Instead of being a perfect sphere, the Earth is an oblate spheroid. Newton based certain of his calculations on a measurement of the distance between Paris and Amiens, France, made by Jean Picard (1620–82), a Jesuit priest and a distinguished astronomer. In 1655 Picard was appointed professor of astronomy at the Collège de France. Picard agreed with Newton, that the Earth is oblate—flattened at the poles.

Another problem exercising astronomers at the time was how to measure longitude accurately, and Picard corresponded on this topic with the professor of astronomy at the University of Bologna, in Italy, Giovanni Domenico Cassini (1625–1712). In 1669 Picard persuaded the French king, Louis XIV, to invite Cassini to become director of the Paris Observatory. Cassini moved to Paris with the blessing of his patron, Pope Clement IX. Four years later, in 1673, the king awarded French citizenship to Cassini to make sure he remained in France. Cassini's first names are therefore often given as Jean-Dominique, and he is often described as a French astronomer and mathematician.

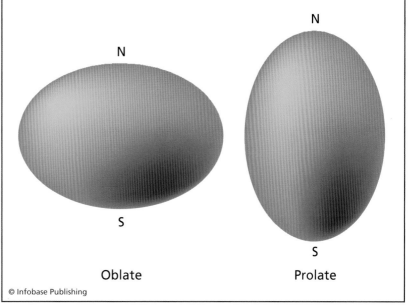

© Infobase Publishing

Oblate and prolate

Picard found Cassini difficult to work with. Although basically quiet and mild mannered, Cassini was also stubborn and opposed many of the changes Picard wished to make at the observatory. The shape of the Earth was one of the matters on which they disagreed. Picard thought it oblate; Cassini agreed that it was flattened but thought it was flattened at the equator, in other words, that it was prolate, like a rather fat egg. The diagram shows the difference.

Happily, Newton had provided a means by which the matter could be resolved. His study of gravity had shown that the force of gravitational attraction between two bodies decreases in proportion to the square of the distance between them, which is expressed mathematically as $F = (Gm_1m_2)/d^2$, where F is the force of gravity between the two bodies, G is a constant, m_1 and m_2 are their respective masses, and d is the distance between them. This is known as the *inverse square law.* Gravity draws objects toward the center of the body, so if the Earth is flattened, the force of gravity measured at its surface should be less where the Earth's diameter is greatest and greatest where the diameter is least. All the scientists had to do was measure the force of gravity somewhere near the North Pole and somewhere near the equator and compare the two findings. The strength of gravity at Paris was well established, and Paris is far enough north for the purpose of the experiment, so only one measurement was required.

Accordingly, in 1672 an expedition led by the astronomer and mathematician Jean Richer (1630–96) and including Picard sailed to Cayenne, in French Guiana. They took with them a clock made by the Dutch physicist and astronomer Christiaan Huygens (1629–95).

The principle of gravity measurement was simple. Huygens had invented the pendulum clock, which is based on the discovery made by Galileo Galilei (1564–1642) that a pendulum will swing in a completely regular way provided some mechanism is provided to prevent it from slowing due to friction and air resistance. Huygens devised a system of weights to provide just the right amount of power and gearing to translate the motion of the pendulum into the movement of hands on a clock face. A pendulum swings under the force of gravity. If that force increases, the pendulum will swing faster and the clock will gain time, and if gravity decreases, it will swing more slowly and the clock will lose time. Pendulum clocks are extremely sensitive. They are adjusted to run at the correct speed by making fine alterations to the length of the pendulum.

The scientists also needed a means of checking the time shown by their clock with the real time. Obviously, they could not take two

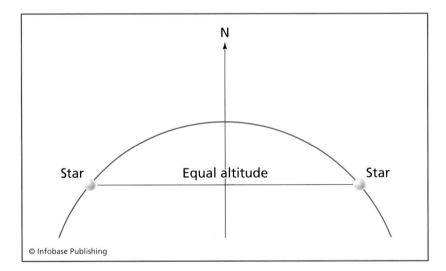

© Infobase Publishing

Determining the meridian by the "equal altitude" method. A quadrant is pointed to the east and set to an arbitrary angle of declination. When the chosen star crosses that angle, the time is noted and the quadrant turned to the west, leaving the angle unaltered. When the star crosses that angle, the time is noted. The time halfway between the observations is local noon.

similar clocks, as a change in gravity that affected one would affect both. Provided the sky is clear, it is very simple to measure local noon, however. The moment when the Sun reaches its highest point in the sky is local noon. If the clock does not show noon, then it is the clock that is wrong by the length of time the clock shows before or after noon. In order to determine noon, the observer aligns an instrument such as a quadrant (see illustration, page 6) with true north and south. This is done with reference to the stars, so it must be determined at night and recorded, for example, by a line drawn on the ground. The instrument is then aligned to point to the south in the Northern Hemisphere and to the north in the Southern Hemisphere. A line from the horizon directly ahead of the observer and to the north or south, passing directly over the observer's head, and ending at the horizon directly to the rear is called a *meridian*. A part of the meridian line, measured on the Earth's surface, is called a *meridian arc*. The Sun and all the stars will pass from east to west, and the Sun will cross the meridian at noon. It is difficult to measure precisely the point at which a celestial object crosses the meridian, so surveyors often use the "equal altitude" method, illustrated in the diagram. The instrument is set to a particular declination to the east of the meridian. When the body crosses that declination the time is noted, and the instrument is moved to the west of the meridian and the time noted when the body passes that point. The time halfway between the two times is the time when the body crossed the meridian.

Richer and Picard calibrated a Huygens clock very carefully before departing. When they conducted their experiment at Cayenne they found that the clock was losing 2.5 minutes a day and to correct it so it showed the correct time they had to lengthen the pendulum. Picard concluded that Cayenne is farther from the center of the Earth than Paris is and, therefore, the Earth is oblate. Cassini would not accept the result, maintaining that the measurements had not been made properly. A second expedition was dispatched in 1681 to Gorée, an island off Senegal, West Africa, and this time the scientists were trained by Cassini himself, to make sure they did not repeat what he thought were Picard's mistakes. This time they used a method Cassini had devised to determine the meridian, and they took two long-case pendulum clocks, calibrated in Paris, one to run on *mean time* (local Sun-based time) and the other on *sidereal time* (star-based time).

Both their clocks ran slow, and they had to lengthen their pendulums to correct them. This time Cassini blamed the clocks, but really he had lost the argument. Picard was right: The Earth is oblate.

ANAXIMANDER—AND THE FIRST MAP

Mapmakers usually draw their maps on paper or animal skin. These are perishable materials that seldom survive for very long. Although the ancient Greeks and Romans must have possessed and used maps, very few of them have been found. Historians can only guess at what they may have been like from the few written descriptions of them that have survived. Eratosthenes drew a map that became famous, but his map was not the first. Historians believe that the first cartographer was Anaximander (611–547 B.C.E.).

Anaximander was born and died at Miletus, a city on the Mediterranean coast of what is now Turkey. He was a pupil of Thales (ca. 640–546 B.C.E.), who was also a native of Miletus. Centuries later the Greeks credited Thales with having founded Greek science, mathematics, philosophy, and in fact just about every conceivable branch of knowledge. This may be exaggeration, but certainly Thales founded the Milesian school of learning, and Anaximander was its second most important philosopher and a truly original thinker.

Until the Greek inventor Ctesibius, who flourished in Alexandria in the second century B.C.E., constructed an improved version of the *clepsydra* (water clock) invented by the ancient Egyptians, the sundial was the most accurate device for measuring the passage of time. The Egyptians and Babylonians had used sundials for centuries, but Anaximander introduced them to the Greeks, using a version with a vertical needle, called a gnomon. He measured the changing length and angle of the shadow cast by the gnomon to determine the dates of the *equinox*es and *solstice*s, which allowed him to calculate the length of the seasons. He also estimated the size of the Sun. Various authorities report that he thought it was the same size as Earth, or 27 times bigger, or 28 times bigger.

Anaximander recognized that all the visible stars rotate around the Pole Star (Polaris), which suggested to him that the universe is spherical, rather than being a hemisphere over the Earth. He also noted that the stars changed their positions in the sky when he traveled north or south. This led him to conclude that the surface of the

Earth is curved, but only in a north-south direction. If the Earth is curved only in one direction its shape must be cylindrical, and Anaximander believed the Earth is a cylinder, three times longer than it was high, with its axis running east-west, and floating freely in space (unsupported by elephants, turtles, columns of water, or any other being or structure)—a view that immediately raised the question of why the Earth does not fall, which until Newton no one could answer satisfactorily. It was not until the Earth was first photographed from space that everyone could see that the planet floats unsupported, yet Anaximander worked out for himself that this must be so.

If the Earth is cylindrical, it is tempting to suppose that the surface on which people live is either the outside of the cylinder or perhaps the inside. That is not the way Anaximander saw it. His cylinder is more like a drum, and people live on its top. That image is the one he sought to record in his map of the entire known world. His map

A reconstruction of Anaximander's world map

has not survived, but the Greek historian Herodotus, who lived in the fifth century B.C.E., saw and described maps like it and probably derived from it. The illustration shows how it may have looked. The map is circular, because it depicts the top of a cylinder, and the river, called Ocean, surrounds all the land. The Mediterranean Sea lies at the center of the map—and of the world, of course, so far as Anaximander was concerned. The world is divided into two halves by a line (not shown on this reconstruction of the map) passing through Delphi. The Greeks believed that Delphi, northeast of

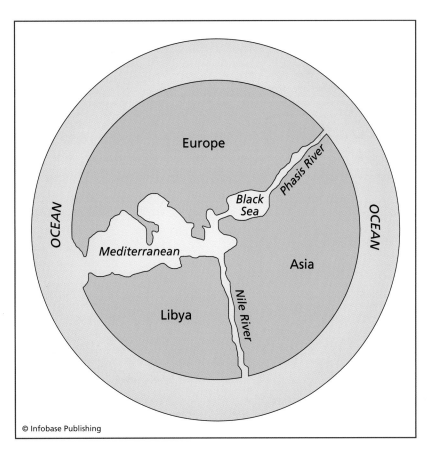

© Infobase Publishing

Athens near Mount Olympus, was the world's navel. Europe lies to the north of the dividing line, and north of the Black Sea and Phasis River. Europe contains Spain, Italy, and Greece. Libya and Asia lie to the south, separated by the Nile. Asia includes Palestine, Assyria, Persia, and Arabia. Libya includes Egypt. The habitable part of the world, from the Greek point of view, consisted of the fairly narrow strips of land to the north and south of the Mediterranean. Farther north were cold lands inhabited by mythical peoples, and to the south were hot lands, inhabited by people who had been burned and were black.

Although Anaximander drew the first map of the entire world, fragments of earlier maps made on clay tablets have survived. One, from excavations at the Mesopotamian city of Nuzu (now Yorghan Tepe in northern Iraq) was made in about 2300 B.C.E. It shows an area of land surrounded by hills on two sides and divided by a waterway (a river or perhaps canal). Cuneiform writing in the center gives the name of the owner of the land and the size of his holding (about 30 acres [12 ha]), and *north, south, east,* and *west* are written on the four sides. The map is drawn with north to the left. This is the earliest known use of the cardinal points of the compass and the earliest known map.

In about 600 B.C.E., during the lifetime of Anaximander, a Mesopotamian scholar inscribed a clay tablet with a schematic map of the entire world known to Babylonians. Their world is surrounded by an ocean, with the four regions Babylonians believed existed at the edge of the world represented by triangles jutting into it. The map shows the Euphrates River crossed by a rectangular shape labeled *Babylon* and leading to a larger rectangle representing the marshes of southern Iraq, beside the Persian Gulf. Various neighboring countries are shown as small circles to either side of Babylonia.

HECATAEUS—AND THE FLAT EARTH SURROUNDED BY AN OCEAN

Miletus was one of 12 Greek cities located on the coast of Asia Minor (modern Turkey). It lay near the mouth of the Maeander (Menderes) River (from which we derive our word *meander*). The city had an ancient history. According to Homer it existed during the Trojan War, and Hittite histories mention it in about 1320 B.C.E., when it was involved in a rebellion. The city became involved in another war in the eighth century B.C.E. Although the city was Greek, geographically it lay on the edge of the Lydian Empire, ruled by Croesus. Relations between the Greeks and Lydians were harmonious, but when Cyrus the Great of Persia defeated Croesus, Miletus fell under Persian rule. The Greeks defeated the Persians on the Greek mainland in 479 B.C.E. and in 334 B.C.E. Alexander the Great finally freed Miletus from the Persians. The map shows Miletus's location.

The city was laid out on a grid plan that later became the layout the Romans adopted as their basic town plan, and until silt carried down by the Maeander clogged it, Miletus had a harbor. It was an important commercial center and a cosmopolitan city that grew into an important center for learning. Miletus must also have been politi-

Lydia and the location of Miletus

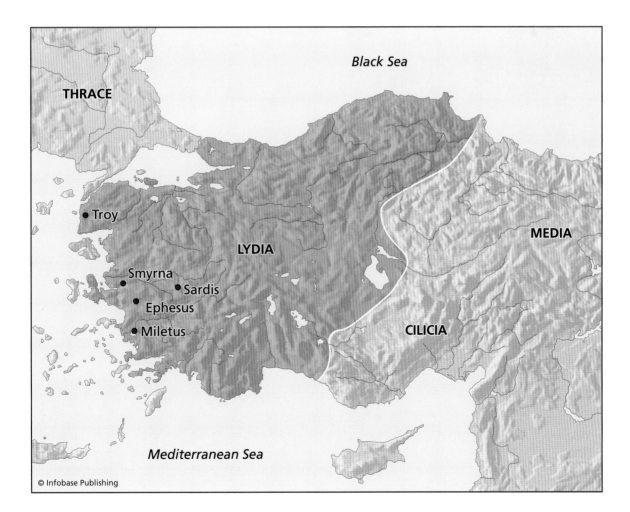

© Infobase Publishing

cally turbulent. It had a large slave population and suffered a bitter struggle between the rich and poor sectors of free society, too. Poor people killed the wives and children of aristocrats, then the aristocrats regained control and killed their opponents. From the time the city was founded, it was ruled by a landowning aristocracy. Gradually rich merchants replaced this ruling class. Then a democratic party supported a leader who overthrew the merchants and became a tyrant. Nowadays the word *tyrant* is defined as a cruel, oppressive ruler, but in ancient Greece it was simply a person who seized power illegally and then ruled without reference to any other authority. A tyrant was not necessarily cruel. Milesian politics were typical of Greek cities at the time. Miletus was a lively and sometimes violent place, but it was not unusual.

Miletus is the city where Hecataeus (ca. 550–ca. 476 B.C.E.) was born into a wealthy family. It is also where he spent much of his life and where he died. A follower of the Milesian school founded by Thales, Hecataeus was a geographer and a historian. He based his ideas on the world he saw around him and had no time for rumors and travelers' tales. He wrote a history of the Greek heroes, called *Genealogiai*, of which a few fragments have survived, and in one of these he states: "I write down what I deem true, for the stories of the Greeks are manifold and seem to me ridiculous."

Before settling down, Hecataeus traveled widely through the Persian Empire and visited Egypt, which had recently come under Persian dominion. Hecataeus described his travels in two volumes called *Ges Periodos* (meaning "Travels around the Earth" or "Description of the Earth"). The work was a survey of the Mediterranean coast. In the first volume he describes Europe by region, from east to west, and in the second he describes Asia, working from east to west and including North Africa. Sometimes his narrative leaves the coast to follow a major river. He was interested in everything he encountered—people, plants, animals, mountains, rivers, distances, and cities, as well as the stories and myths that he heard. He did not criticize the beliefs he considered absurd, simply recording them and in doing so exposing the ways they contradicted each other. *Ges Periodos* survives as rather more than 370 fragments, most of them very short, but one of the fragments includes a map depicting the world Hecataeus had explored. The map summarized his written description.

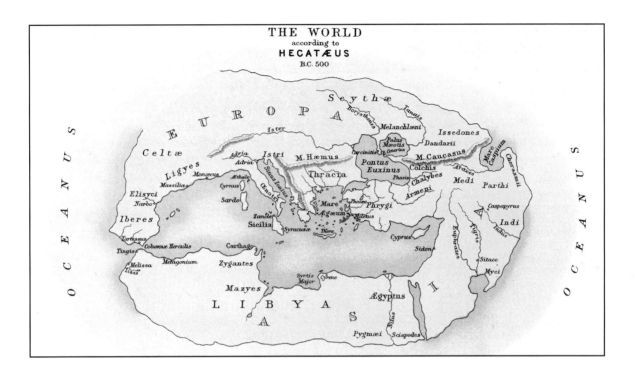

The Hecataeus map is clearly derived from the map of Anaximander, but Hecataeus had corrected the earlier map and added more detail. The illustration shows what the Hecataeus map probably looked like. Like Anaximander, Hecataeus shows the known world as circular and surrounded by the Oceanus River. Greece is at the center, with the Mediterranean Sea extending a little way eastward and westward all the way to the edge. The Black and Caspian Seas extend the line of the Mediterranean to the east and the Red Sea runs south. Europe (Europa) lies to the north and Asia to the south. The map identifies islands and countries and the names of the peoples inhabiting certain areas. Hecataeus is the first geographer to mention the Celtae (Celts) of northwestern Europe.

Hecataeus's map is partly schematic and partly an attempt at a true representation. Compared with a modern map, however, one important feature is missing: There are no grid lines. Modern maps are divided by lines of latitude and longitude or by grid lines counted from an origin determined by the cartographer. Grid lines allow the positions of places to be reported accurately by means of a simple system of coordinates, and they allow mapmakers to plot

The map of Hecataeus of Miletus *(Granger Collection)*

features correctly. Grid coordinates still lay far in the future, however. Cartographers would have to wait some 350 years before they were introduced by Hipparchus, one of the greatest astronomers and mathematicians who ever lived (see sidebar below).

MARCUS AGRIPPA—AND THE PEUTINGER TABLE

The Milesian geographers debated the shape and size of the world and constructed maps to show the relative locations of places known to them, and they did so in the spirit of scientific study. They were exploring the world around them. Several centuries later the Romans were busy conquering the known world and assembling a vast empire. They also needed maps but for a more practical purpose. The Roman authorities needed to move troops, administrators, tax collectors, and other officials to distant provinces. Roman maps

HIPPARCHUS—AND HOW *LATITUDE* AND *LONGITUDE* ACQUIRED THEIR NAMES

Modern maps show land and sea overlaid by a grid of lines, one set running east and west and the other running north and south. These are lines of latitude and longitude, and they make it possible to specify the location of any place on the map—or anywhere in the world—by means of two coordinates, nowadays reported in degrees, minutes, and seconds. Because it does not rely on landmarks, the method works just as well for reporting the position of ships at sea as it does for places on land. It also improves the accuracy of the map, because a degree of latitude or longitude has a precise length, so the angular distance between points—the distance measured in degrees—is easily translated into linear distance, for example, in miles or kilometers.

The names are Latin, from *latus*, "broad," and *longus*, "long," but the first person to draw a line of latitude on his map was the Greek geographer Dicaearchus (ca. 355–ca. 285 B.C.E.), who lived in Messina, Sicily. His line linked the places where on any given day in the year the noonday Sun was at the same declination, and he developed this into a grid system.

Many years later Hipparchus (ca. 190–ca. 120 B.C.E.) took this idea further. The greatest of all the Greek astronomers, Hipparchus measured the position of the Moon against the stars under changing conditions and used this to calculate the distance between the Earth and Moon as being between 59 and 67 times the radius of the Earth. This is very close to the correct value of 60 Earth radii, and if Hipparchus had contained much detail. They showed roads and landscape features, and early maps showed farms and estates based on land surveys. The Romans were much less concerned than the Greeks with measuring and showing latitude and longitude or, indeed, with making accurate representations of the lands they depicted. These were stylized road maps rather than geographical works. They informed the traveler of what lay ahead and the distance to the next stopover.

The earliest record of a Roman map concerns one of Sardinia drawn in 174 B.C.E., although land surveys were being conducted around 165 B.C.E. to produce what were estate plans, rather than maps. At first the maps were circular, but later maps based on land surveys were square or rectangular, a shape partly dictated by the surveying techniques the Romans used but also favored because it was convenient for placing on the walls of temples and other public buildings.

used Eratosthenes' value for the diameter of the Earth, he would have found that the Moon is about 250,000 miles (402,250 km) distant. He also calculated the length of the year and was accurate to within 6.5 minutes, and he discovered the *precession of the equinoxes,* which is the way the Earth's position in its orbit at the equinoxes changes over time. Hipparchus calculated the rate of precession as 46" (seconds of arc) a year; the true value is 50.27". In his study of angles and the orbits of the Sun and Moon, Hipparchus was the first Greek astronomer to use the Babylonian system of dividing a circle into 360° each of 60' (arc minutes), and he is considered to have invented trigonometry.

Hipparchus used astronomical observations to construct a grid of 360° of latitude and longitude, which he imposed on a map of the world, and criticized Eratosthenes for basing his map on the positions of places he had accepted from others, which introduced inaccuracies. Astronomers and surveyors now use extremely accurate instruments to make precise measurements of latitude and longitude. Such instruments did not exist in Hipparchus's day, so his measurements were inevitably approximate. Nevertheless, it is to him that we owe both the method of imposing a grid on maps and the calibration of the grid into degrees, minutes, and seconds of arc, with 360° making a circle.

Hipparchus was born at Nicaea, in Bythnia (now Iznik, Turkey). Modern astronomers have calculated that when he compiled his star catalog showing the positions of the brightest stars, Hipparchus must have been located at 36°15'N. This is the latitude of Rhodes, Greece, and it is assumed that he spent his final years at Rhodes and died there.

Shortly before his death Julius Caesar (100–44 B.C.E.) began to compile a map of the lands he had conquered. According to late Roman and medieval historians, Caesar employed four Greek geographers (probably former slaves who had gained their freedom and were living in Rome). The Greek cartographers began work in 44 B.C.E., the year in which Julius Caesar was assassinated, and all work ceased with his death. Nothing is known of the way the map was organized or even whether work on it began before Caesar died, but his dream of a map of the empire survived. That map was made by the statesman and general Marcus Agrippa.

Marcus Vipsanius Agrippa (63–12 B.C.E.) was born into a wealthy, high-status family. Gaius Octavius (63 B.C.E.–14 C.E.) was a boyhood friend, and they both served under Julius Caesar as officers in the Roman cavalry, fighting in the Battle of Munda in 45 B.C.E. Atia, Octavius's mother, was the daughter of Julia, who was Julius Caesar's sister, and Caesar regarded Octavius as the most able of all his male relatives. After the army returned to Rome, Caesar adopted Octavius as his son. When Caesar was killed, Octavius was away studying in Illyria (modern Albania). On hearing the news, and the news that he was Caesar's heir, he hurried back to Rome. At that time he was only 18 years old, and his position was weak, but he entered Roman politics. There were wars with rivals, and Agrippa fought alongside the emperor as his most senior general, but it was not until 30 B.C.E. that Octavius became undisputed emperor following the capture of Egypt and the suicide of Mark Antony (Marcus Antonius; ca. 82–30 B.C.E.) and Cleopatra (69–30 B.C.E.). Octavius immediately began to institute reforms, aided by his friend Agrippa, and in 27 B.C.E. the Roman senate conferred on him the imperial title *Augustus*.

Agrippa's influence grew with the rise of Augustus, and as well as being a highly skilled general and admiral, he was very active in public life. He was keenly interested in architecture and designed and supervised the building of the Pantheon, then went on to build and repair aqueducts, construct bathhouses, and lay out gardens. He also sponsored art exhibitions.

In addition to these skills, Agrippa was a geographer and wrote works on geography. The map Julius Caesar had dreamed of was commissioned by Augustus and drawn under Agrippa's supervision. The map of the whole Roman Empire was an immense work that took a large team of surveyors almost 20 years. The map showed more than 50,000 miles (80,000 km) of paved roads, with distances in Roman miles (1 Roman mile = 0.944 mile = 1.52 km) indicated by marking the positions of milestones. The road network formed the skeleton of the map, to which the surveyors added provincial boundaries, rivers, towns and cities, and facilities for travelers, such as temples and baths. Agrippa died before the map was finished, and Augustus took over the supervision himself. When at last it was done, a large version of it was carved in marble and erected on the wall of a portico, the Porticus Vipsaniae (Agrippa's middle name), close to the Forum. Agrippa's map was a political statement. It showed the extent and might of Roman power, and at the same time it suggested that Rome's emperor was the benevolent head of this vast community.

The map covered the area from the southeastern tip of England and the Pyrenees to India, Sri Lanka (called Insula Trapobane), and China and the coast of the imagined eastern ocean in the east, as well as North Africa. It had 534 illustrations, but it greatly distorted shapes. The Nile appeared to flow east-west, and the Mediterranean and Black Seas were long and narrow, looking rather like canals.

The master map has not survived. Neither have the many copies that were made of it, drawn on papyrus scrolls and distributed to administrators and military commanders throughout the empire. But later copies were made, and one of these has survived. It was drawn first in the fourth century C.E. and copied again by a monk in the 13th century. The map was then lost and rediscovered in a library in Worms, Germany, by Conrad Celtes (1459–1508), a German scholar. Celtes was unable to publish the map before his death, so he bequeathed it to a friend, Konrad Peutinger (1465–1547), an antiquarian. It is now known as the Tabula Peutingeriana, or Peutinger Table. The Peutinger family kept it until 1714, and in 1737 it was purchased by the Austrian emperor and placed in the Imperial Court Library. It is now preserved in the Österreichische Nationalbibliothek, in Vienna.

The Peutinger Table is a parchment scroll that is 13 inches (0.34 m) wide and 22 feet (6.7 m) long and designed to be folded into a portfolio. It is made from 11 sections, but a 12th section, showing the British Isles and Iberian Peninsula, is missing.

WILLEBRORD SNELL—AND THE DISCOVERY OF TRIANGULATION

If a map is to be a useful tool for navigation, it must represent distances and the relative locations of features as a person might find them on the ground. It must be accurate. Achieving accuracy is difficult, but the invention of latitude and longitude was a great help. Since these lines are located with reference to their angle from the center of the Earth, they can be drawn on a surface in a standard, unambiguous fashion. Once the grid is in place, the cartographer need only place coastlines, rivers, roads, towns, and other features in their correct positions measured as their latitude and longitude. In order to do that the cartographer must go out into the landscape and measure them.

Unfortunately measuring the distance between two places miles apart is not so simple as it may seem. The most obvious method is to use a measuring device; a chain is the simplest. That works well for small distances across level ground, but most ground is uneven, and the chain will need to pass over all the bumps and depressions. This also raises another question: should the chain measure the actual distance across the surface—the distance a person would walk—or is it to measure only horizontal distance, ignoring vertical distances or finding some other way to show them? In fact, maps show only horizontal distances, using some combination of contour lines and shading to suggest elevation. This is necessary, because a map based on measurements along the ground surface would distort areas.

There is another problem. Metals expand when their temperature rises and contract when it falls, so the length of a metal chain changes with the temperature. The extent of the change may be too small to matter for some purposes, but it is unacceptable when surveying for a map.

A wooden or glass rod is an alternative to a chain that avoids the problem of contraction and expansion. The rod must be cut to a very precise length then laid in a wooden trough supported on legs or a stand. The equipment is cumbersome, but rods have been used successfully. Chains can also be used, provided their length is checked before measuring commences, and checked again if there should be a change in temperature before work is completed. Chains, too, are often laid in troughs for greater accuracy.

Even so, this is a cumbersome way to measure long distances, and there are far too many opportunities for error for it to be much use to mapmakers, especially since there is a much quicker and more accurate method—triangulation. Tycho Brahe (1546–1601), the Danish astronomer, is said to have tried this surveying technique and so did the Dutch mathematician and geographer Regnier Gemma Frisius (1508–55), but the name most closely associated with developing and applying the method is that of Snell, also known by the Latinized version of his name, Snellius.

Willebrord van Roijen (or Royen) Snell (1580–1626) was born and died in Leiden, Netherlands. Although he studied law at the University of Leiden, he was most interested in mathematics and succeeded his father as professor of mathematics at the university. He is famous for having explained the refraction of light, discovering the sine law, and improving the method for calculating approximate values for π. Snell also founded the scientific discipline of *geodesy,* which is the measuring, surveying, and mapping of the Earth's surface. In 1617 he published *Eratosthenes Batavus,* a book about Eratosthenes, in which he also described the principles of triangulation.

Triangulation is a trigonometric method for measuring distances. At some stage, but not necessarily before commencing the rest of the task, the surveyor must measure the length of a baseline. This must be done as precisely as possible, and surveyors often measure a second baseline as a check. The baseline can be located on level ground, or even across the surface of water, provided the workers can lay their measuring chain or rods across it.

The work then involves measuring the angles between both ends of the baseline(s) and landmarks within the surveyor's line of sight and between each landmark and others. If real landmarks exist they can be used, but the surveyors often erect towers to act as triangulation points. The angles are recorded and when plotted on a chart they construct a series of triangles, as illustrated in the diagram. If the length of one side and the angles of a triangle are known, it is possible to calculate the length of the remaining two sides by trigonometry. When the surveying is complete, a mathematician solves all of the triangles. It is then possible to calculate the length of any line inside the area they cover.

In 1669 Picard began the task of measuring the size of the Earth on behalf of France's Académie Royale, and he used Snell's triangulation. The first step was to measure the length of one degree of latitude along a meridian arc. A meridian arc is an imaginary line drawn on

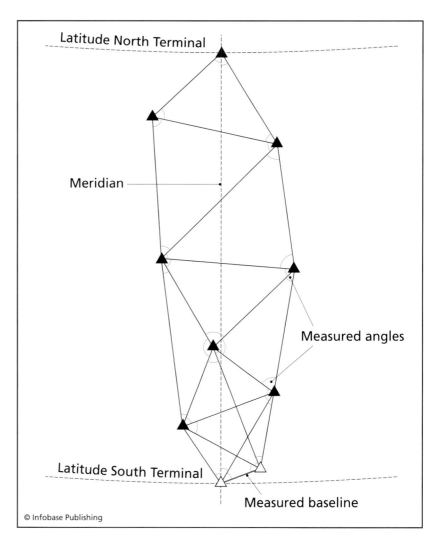

Latitude North Terminal

Meridian

Measured angles

Latitude South Terminal

Measured baseline

© Infobase Publishing

Using triangulation to measure one degree of latitude

the ground and running north and south (see "Oblate or Prolate," pages 14–18). It is not necessary to measure the whole of one degree; a smaller length will suffice. Picard chose two points on the meridian arc and determined their latitude astronomically, measured a chain of triangles between them, and calculated that the length of one degree of latitude in France is 69.16 miles (111.3 km). Because the Earth is oblate, however, the length of a degree of latitude is not everywhere the same: It is shorter close to the equator than it is close to the North and South Poles.

THE WEIGHT OF MOUNTAINS

Schiehallion (the name is from the Gaelic for "fairy hill of the Caledonians") is a mountain 3,554 feet (1,083 m) high that stands alone in the Highlands of Scotland. It is popular with climbers and hill walkers, but it has an honorable claim to scientific fame. It is the first mountain ever to have been weighed. For enthusiasts of trivia, it is also the geographic center of Scotland.

Surveyors measuring meridian arcs in different places had found an anomaly. Sometimes the measurements were clearly incorrect but only by very small amounts. Newton had suggested that since all bodies exert a gravitational attraction, mountains must also do so, and scientists suspected that the errors in their measurements were due to the gravitational attraction of nearby mountains.

Errors could arise because surveying instruments used to measure latitude must be absolutely upright. The declination of a star, and therefore the latitude, is measured along a line between the star and the center of the Earth, and the instrument must be upright. To ensure that it is, the instrument is fitted with a plumb line and is set up so that the plumb line hangs vertically (see the drawing of a quadrant on page 6).

Surveyors and cartographers use two shapes for the Earth. The *spheroid,* or *ellipsoid,* is the best fit for the actual shape of the Earth. The *geoid* is the actual shape of the Earth; gravity acts at right angles to any point on the surface of the geoid. If the entire Earth were covered by oceans, and no wind or current stirred the surface, then that surface would represent the geoid. The mass of a mountain will deflect a plumb line—the weight is drawn toward the mountain—and a reading of latitude will be slightly different as a consequence. The question was, By how much? Some scientists suggested that 10 arc seconds was the very least deviation possible. That would amount to more than 1,000 feet (300 m) on the Earth's surface. Mapmakers needed to measure the deviation because the surveying results found it to be significantly smaller than Newton had predicted. The eminent physicist Henry Cavendish (1731–1810) suggested a reason for the difference. He said that ocean basins might exert an opposite effect. The two effects together came to be called the "attraction of mountains and the deficiency of oceans."

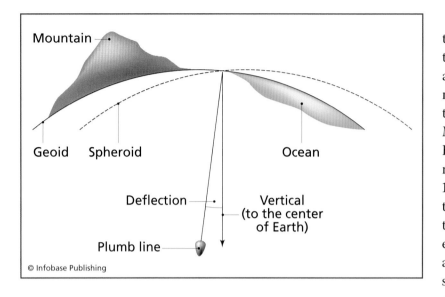

© Infobase Publishing

The effect of gravity anomalies. The greater mass of the continental crust compared to the ocean basin deflects the plumb line.

The task of measuring the gravitational attraction of Schiehallion was assigned to the astronomer royal to George III, the Reverend Nevil David Maskelyne (1731–1811). He applied for royal permission and funding in 1772, and in 1774 he set to work at the head of a team of highly skilled and experienced surveyors and astronomers. They measured the gravitational effect on the pendulums of the most accurate clocks in the world, and in 1775 Maskelyne reported to the Royal Society that Schiehallion would deflect the plumb line of a measuring instrument by 11.66 arc seconds.

The Royal Society had appointed Charles Hutton (1737–1823), foreign secretary to the society and professor of mathematics at the Royal Military Academy, Woolwich, to perform all the calculations needed to produce a result from Maskelyne's data, and Hutton went on to calculate the mass of the mountain and, from that, the density of the Earth. He concluded that Schiehallion was made from a solid mass of rock. He reasoned that rock is 2.5 times denser than water and therefore that the Earth is 4.5 times denser than water. The Earth's density is now known to be a little more than 5.5 times that of water.

As well as contributing to the accurate mapping of the world, the Schiehallion experiment also resolved another puzzle. Until then no one had been entirely certain whether the Earth was solid or hollow. Schiehallion confirmed that it is solid.

PTOLEMY—AND HOW TO REPRESENT A SPHERE ON A FLAT SURFACE

Roman maps were designed as administrative and military tools. They needed to show administrators and commanders the best routes and traveling times between important places and were based on the reported experiences of people who had walked those routes. The maps were not derived from any underlying scientific theory. Scientific studies still continued, however, and the focus for them shifted to Alexandria, a cosmopolitan city and the center of the Greek administration of Egypt, where they reached their peak in the work of Ptolemy.

Claudius Ptolemaeus, usually known simply as Ptolemy, flourished between about 127 and 151 C.E. Very little is known about his life. Arabic scholars reported that he was 78 years old when he died. No one is certain whether he was Greek or Egyptian, but he was probably born in the Greek city of Ptolemais Hermii, beside the Nile in Upper Egypt. Historians believe his name refers to where he was born and that he was not related to the Ptolemies who ruled Egypt. He is always described as being Greek. Ptolemy is most famous as an astronomer and mathematician who drew together ideas and information from many sources, in particular building on the work of Hipparchus. He recorded his astronomical and mathematical work in a book called *Mathematike Syntaxis* (Mathematical collection), which came to be known informally as *ho megas astronomas* (the great astronomer) to distinguish it from another work called the little astronomer. Arabic astronomers who translated it in the ninth century called it "the greatest" (*megiste* in Greek), which became corrupted to *Almagest*, the name by which the work has been known ever since.

Ptolemy was also a geographer and wrote *Geographike Huphegesis* (*Geography*), an eight-volume work. In book 1, Ptolemy discussed the problem every mapmaker faces in trying to depict more than a fairly small local area: What is the best way to depict the surface of a sphere in a two-dimensional map? The other seven volumes were little more than lists of about 8,000 place-names with details of their location.

Ptolemy drew a map of the known world, showing part of Europe, Asia, and Africa, with the Mediterranean Sea and Indian Ocean. His original map has not survived, but it was reconstructed from his description in the *Geography* and printed at Ulm, Germany, in 1472. That is the version reproduced here in the figure. The heads around the map show the winds of the world. In addition, the first volume contained an atlas with 10 maps of different parts of Europe, 12 of Asia, and four of Africa.

His astronomical experience made Ptolemy ideally suited to follow the method originally devised by Hipparchus. He divided the equator into 360 degrees and drew other lines of latitude on either side of it. His meridians (lines of longitude) were *great circles* that crossed the equator at right angles. A great circle is the shorter arc of a circle drawn on the surface of a sphere the center of which is also the center of the sphere. His result was a curved map, representing a section of the spherical Earth.

The method was reliable, but Ptolemy was unable to apply it correctly. He had a few astronomical measurements to guide him and relied mainly on reports from travelers, which amounted to a kind of *dead reckoning*—estimating the location of a place from the time taken to reach it, moving at a known speed, from another place whose location is known. This is accurate enough for approximate navigation but not for drawing reliable maps. Ptolemy rejected Eratosthenes' estimate of the Earth's circumference in favor of Poseidonius's more recent calculation. Poseidonius arrived at a figure 30 percent smaller than the one Eratosthenes had calculated. It was an under-

Ptolemy's world map *(Sheila Terry/Science Photo Library)*

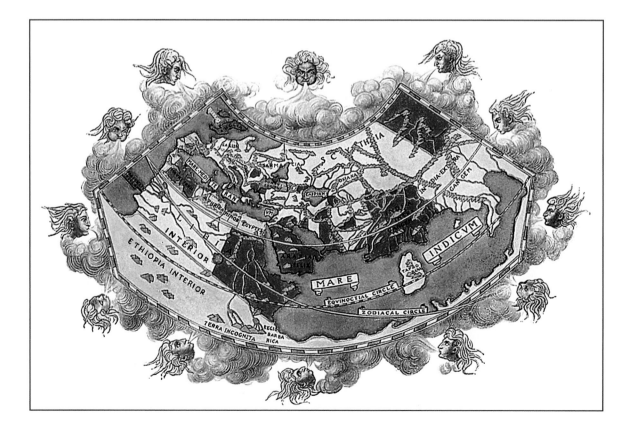

standable decision, but unfortunately Poseidonius was mistaken (see "Poseidonius—and Why Columbus Thought He Had Reached Japan," pages 9–12), and the distances were wrong by that amount on Ptolemy's map. Ptolemy placed the equator too far north, and Rhodes is the only place shown in its correct latitude. Europe extends through 130° of longitude, but Ptolemy makes it 180° wide. He shows Scotland extending eastward at right angles to England, the Sea of Azov is too far north and much too big, the Persian Gulf is too wide, the Indian Ocean is enclosed, India is too small, and Sri Lanka is too large.

It is easy to criticize Ptolemy, but his maps represented a major advance. He sought to depict the world as it really is and not as myths and legends would have it, and he developed a properly scientific method for setting about the task. His work ran into difficulty for lack of data. He could not travel the world taking measurements or employ a large team of trained astronomer-surveyors to do so. His world map was full of errors, but it was a serious and bold attempt at accuracy and much better than the maps produced in medieval Europe.

MEDIEVAL MAPS

In the sixth century, a merchant and traveler called Cosmas of Alexandria sailed around the shores of the Red Sea and Indian Ocean. He is surnamed Indicopleustes, a name meaning "the Indian navigator," although he never reached India.

Cosmas was also a theologian who became a monk in about 548, and some time between 535 and 547 he wrote *Topographia Christiana* (*Christian Topography*), a book in which he sought to prove the literal accuracy of the biblical description of the universe. In particular Cosmas denounced the false and heathen idea that the world is spherical and that there are people living on the far side of it, in the Antipodes. He maintained that the Tabernacle of Moses is a representation of the universe, and consequently an accurate map of the world should show its parts arranged like the parts of the Tabernacle. His resulting map shows the Earth as rectangular and flat. Above it is the firmament—the sky with all the stars—a vaulted roof, and heaven lies above the firmament. The inhabited part of the Earth is at the center of the rectangular plane, surrounded by an ocean, and on the far side of the ocean there is the paradise of Adam. The Sun is much smaller than the Earth, and it moves around a conical mountain in the north, disappearing behind the mountain at night. It orbits the peak in summer and the base in winter, thus explaining the difference in day length between summer and winter.

Cosmas had drawn a map in which the actual arrangement of coastlines, continents, islands, and rivers was of minor importance. What mattered were the relative locations of the inhabited and spiritual worlds. The map had to show where paradise and heaven are to be found. The map is a guide to a spiritual, not a physical, journey.

This view of the world influenced maps throughout the Middle Ages. Few scholars went as far as Cosmas in denying the spherical shape of the Earth, and the Church never did so, but the Earth appears as a flat disk in their maps. The disk is invariably surrounded by an ocean, which is a very ancient idea. Jerusalem is usually placed at the center of the world map, and east is often at the top.

This general plan developed into "T and O" maps, sometimes called Beatine or Beatus maps because one of the earliest of them was drawn in the eighth century by the Spanish monk Beatus of Liébana. The earliest printed example, reproduced in the illustration, was made in Augsburg, Germany, in 1472 by Guntherus Ziner and appeared in the *Etymologiae of Isidore of Seville*. Like all maps of this type, it shows the disk-shaped world—"'round' like the roundness of a circle, because it is like a wheel," in the original description—surrounded by the ocean (Mare-Oceanum), the "O." East (Oriens) is at the top. The map shows the continents of Asia, Europe, and

The earliest printed T and O map, made in 1472, by Guntherus Ziner

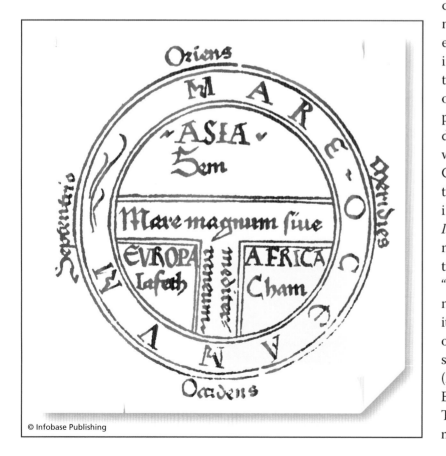

© Infobase Publishing

Africa. The Mediterranean Sea forms the shaft of the *"T,"* separating Europe and Africa, and another sea forms the bar of the *T;* the Nile running horizontally divides Asia from Europe and Africa in most of these maps. This version allots the continents to the sons of Noah, Shem (*Sem* on the map), Japheth (*Iafeth*), and Ham (*Cham*).

T and O maps were very versatile. Their basic design was so simple that one could be drawn small enough to fit inside the letter *O* or even the top of a *P* in an illuminated manuscript. Alternatively, it could be huge. One of the most famous examples is the Hereford Mappa Mundi, made in about 1290 by Richard de Bello and held at Hereford Cathedral, in England. This world map measures 5.4 × 4.4 feet (1.65 × 1.35 m) and is drawn on vellum (calfskin).

By the 13th century large T and O maps had become very elaborate. Place-names were shown and all kinds of detail appeared in the form of tiny pictures. Some illustrated genuine features, and others were more fanciful, being taken from accounts written by travelers.

Large and detailed or small and simple, all of these maps reflected the worldview of the monks and scholars who drew them. They showed the theological arrangement of the world, and although many of them were very beautiful, they would have been of little use to sailors or to someone wishing to travel from one town to another across open country. Realistic maps began to appear in Italy in the 14th century, a product of Renaissance curiosity about the natural world.

FRA MAURO—AND HIS MAP OF THE WORLD

Mapmakers began catering to the needs of sailors early in the 14th century. That is when portolan charts began to appear. A *portolano*, from the Italian *porto* meaning "port," is a book of sailing directions for the use of mariners, and portolan charts were much more accurate than any of the medieval maps. The first of them may have been drawn in about 1270, but they became widespread in the 14th century and remained in use until about 1600, changing very little. The charts were drawn by Italian mapmakers, and they depicted the Mediterranean and Black Seas and the western Atlantic. Coastlines were drawn accurately, and these maps are instantly identifiable because every chart had several points from where eight or 16 lines radiated in the principal compass directions. The chartmakers aimed to serve sailors who navigated by magnetic compass.

World maps grew more accurate at the same time and reached a pinnacle in the 15th century with the work of Fra Mauro. Fra Mauro (whose birth and death dates are not known), an Italian monk from Murano, an island near Venice, was also a highly skilled cartographer. King Afonso V of Portugal commissioned him to prepare a map of the world. Fra Mauro consulted sailors, sometimes paying for their help, and in April 1459 the map was completed and dispatched to Portugal. With the map the ruler of Venice sent a letter addressed to Afonso's uncle, Prince Henry. Henry (1394–1460) was nicknamed the Navigator because of his interest in exploration and his encouragement of Portuguese voyages of discovery. The letter urged him to continue funding such voyages. Clearly the map was meant to be an accurate depiction of the world rather than a schematic representation based on theology.

Fra Mauro's original map has not survived, but he started to make a copy of it for the Venetian authorities. He died before finishing it, but his colleague Andrea Bianco completed it. This map has survived and is now held in Venice at the Biblioteca Nazionale Marciana. It is drawn on parchment, set in a wooden frame, and is about 6.5 feet (2 m) across.

The map has south at the top and observes the convention of showing the continents surrounded by an ocean, so the world looks like a disk—although Fra Mauro knew that the Earth is spherical. Fra Mauro was familiar with the Ptolemy map but improved greatly on it because he had access to the accounts of explorers who by that time had visited many parts of the world of which Ptolemy can have known nothing. Fra Mauro's map shows Africa fairly accurately, has drawings of an Asian junk, and is the first European map to show the Japanese islands.

Fra Mauro's map was so much more advanced than any other that a medal was struck at the time to commemorate it, bearing the appellation *"geographus incomparabilis."* A lunar crater and rock formation are also named after Fra Mauro.

MARTIN BEHAIM—AND THE OLDEST SURVIVING GLOBE

The simplest solution to the problem of depicting a spherical surface on a flat sheet of paper is to abandon the task entirely and make a globe, on which the locations, shapes, and dimensions of features can be shown accurately. A globe is hardly a practical tool for navigation, of course, but it is very useful for instruction. Portraits of famous navigators and geographers often include a globe. It became traditional, as an instantly recognizable clue to the occupation of the sitter.

It is possible that the ancient Greeks made the first globes, but none has survived. The oldest and most valuable surviving globe was made in 1492 and is held at the National Museum in Nuremberg, Germany. The skin of the globe is drawn on parchment and stretched on a sphere 20 inches (50.7 cm) in diameter. It is decorated with flags and pictures of kings, and it shows considerable geographical detail, including more than 1,000 place-names. The equator is divided into 360 degrees, the signs of the zodiac are shown along the ecliptic, and the Tropics and Arctic and Antarctic Circles are marked.

The globe was made by Martin Behaim, a navigator and geographer born in Nuremberg either in 1436 or 1459. Behaim was engaged in trade from Flanders, which took him to Portugal, where he acquired a reputation as an astronomer and became a member of a council advising the king, John II, on navigation. The authenticity of his scientific reputation as well as the stories of his exploratory voyages and of the scientific instruments he invented or improved are highly controversial. In 1486 Behaim married a daughter of Jobst van Huerter, the leader of a group of Flemings who had been allowed to settle in Fayal, in the Azores. Behaim moved to the Azores, but in 1490 he visited Nuremberg. That is when he started work on the globe, assisted by the painter Georg Albrecht Glockenthon. It took two years to complete, and Behaim nicknamed it the Erdapfel (literally, "earth apple," an old German name for a potato). Behaim then returned to Lisbon, where he died on August 8, 1507.

His Erdapfel showed the world the way educated people believed it to be at the time Columbus made his voyage. It relied heavily on Ptolemy's map and was very inaccurate. Its depiction of Africa is quite wrong, and it shows several fabulous islands in the Atlantic. Apart from those the globe shows no land between the Azores and Asia and shows the width of the Atlantic, between Europe and Asia, as 80°, which is the Ptolemaic value and the one Columbus used. Behaim and Columbus were in Portugal at the same time, but there is no evidence that they met.

GERARDUS MERCATOR—AND THE BIRTH OF MODERN MAPS

During the first half of the 16th century a school of cartographers became established in the Netherlands, founded by the astronomer and mathematician Gemma Frisius (1508–55) from Louvain (now in Belgium), who applied himself to many of the problems associated with mapmaking. Gemma also made a pair of globes, one in 1535 and the second in 1537, with the assistance of a young man from the town of Rupelmonde, Flanders, named Gerard de Cramer. De Cramer went on to make two globes of his own, in 1541 and 1551. He continued to work in Louvain until 1552, when he moved to Duisburg, in the Duchy of Cleves (now in Germany). As was the fashion in those times, Gerard de Cramer translated his name into Latin, and it is the Latinized version that became famous. *Cramer* is the Dutch for "merchant," which in Latin is *mercator*. Gerard de Cramer became Gerardus Mercator.

Mercator was born on March 12, 1512, in Gangelt, on the border between the Netherlands and Germany. He was the eighth child of a poor cobbler, and both his parents died while he was still young. His brother Gisbert, a priest living in Rupelmonde, brought him up from that point, and when Gerardus was 16, his brother secured him a place as a scholar (a person holding a scholarship) at the University of Louvain. While he was a student at Louvain, Mercator became attracted to Protestantism. He spent some time traveling in Europe, returning to Louvain in 1534 to study under Gemma Frisius. Mercator learned mapmaking and drew his first maps, of Palestine in 1537 and Flanders in 1540, but in 1544 his Protestantism got him into trouble. He was arrested and imprisoned in Louvain castle. Many of his fellow prisoners were charged with heresy and executed with extreme cruelty, but after seven months Mercator was released and the charges against him were dropped. It was a warning, nevertheless, and that is why he and his family eventually moved to Duisburg, a Calvinist town where he was safe. Mercator prospered in Duisburg, was appointed court cosmographer to Wilhelm of Cleve in 1564, and died there on December 5, 1594.

Mercator established his own business on his arrival in Duisburg in 1552, making globes, maps, and atlases. He was the first person to use the word *atlas* to describe small-scale maps of countries, deriv-

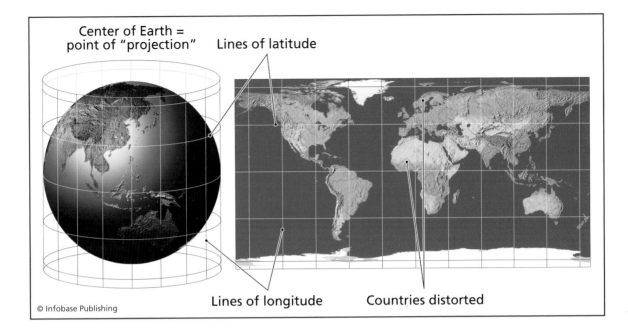

Center of Earth = point of "projection"   Lines of latitude

Lines of longitude   Countries distorted

© Infobase Publishing

ing the word from the Titan Atlas, who carries the world on his shoulders. The secret of Mercator's enduring fame is his solution to the problem of portraying a spherical surface on a two-dimensional surface. Many of the maps published today in atlases of all kinds are based on the Mercator *projection,* which he devised in 1568.

Mercator's is a cylindrical projection. He imagined that a cylinder of paper was wrapped around the globe. Lines drawn from the center of the Earth, through the surface of the sphere, and onto the inside of the cylinder mark the lines of latitude and longitude. The cylinder is then laid flat and the geographical locations of places can be drawn on it, just as though every point had been projected onto the map by a bright lamp inside the sphere and shining through it to cast shadows onto the cylinder. The diagram shows how this is done.

On a Mercator projection the lines of longitude and latitude become straight lines intersecting at right angles. This makes navigation simpler, because if the navigator draws the desired track as a straight line between the starting point and destination, the line will cross all of the meridians at the same angle. A line that does this is called a *rhumb line,* and it allows the navigator to steer a constant course, even though a rhumb line is longer than a great circle. In 1585

A cylindrical map projection. Lines drawn from the center of the Earth pass through lines of latitude and longitude and onto the inside of an imaginary cylinder. When the cylinder is unrolled the positions of land and sea are plotted on the grid.

Mercator published in Amsterdam an atlas of the world that was the most accurate representation of the world up to that time.

There is a problem with the Mercator projection, however. As the diagram shows, the lines of latitude become farther apart with increasing distance from the equator. Just as the difficulty of portraying a sphere does not arise on the small scale of a town or local region because a flat map introduces almost no distortion, so the difficulty with Mercator's projection became serious only when people began exploring in high latitudes. Then it became very serious, because the projection introduced major distortions. A Mercator map greatly exaggerates the size of Greenland and Antarctica and seriously distorts their shapes.

Later cartographers addressed this by developing cylindrical projections in which lines of latitude are equidistant. Some of these are modified versions of Mercator, and when a world map in an atlas is described as "modified Mercator," that is what it means.

Below the Crust

Difficult though it was, drawing a map—essentially a diagram—of the Earth's surface, showing the relative locations of its physical features paled to insignificance compared with the task of finding out what lies below the ground. People walk on the solid surface, quarry rocks from it to construct their buildings, obtain metals, gemstones, and other useful materials from it, but what actually is it? Is the entire Earth solid all the way through? Or is it hollow, perhaps with people living on the inside just as we live on the outside? Why does the Earth sometimes shake, and why are there places where forces tear the rocks apart and send steam, deadly dust, and molten rock streaming forth?

This chapter traces the search for answers to such questions as these. It describes how, little by little and over many centuries, people we would now call geologists assembled the pieces that add up to the modern understanding of the structure of our planet.

IS THE EARTH HOLLOW?

No modern Earth scientist would entertain for a moment the idea that the Earth is a hollow sphere, but it is an idea that has found proponents at various times throughout history, even as recently as the 20th century. The Greeks believed that after they died people entered the realm of Hades. Hades was also the name of the underworld, ruled by Hades and his queen, Persephone, together with her companion, Hecate, and aided by the Erinnyes. These were three winged daughters, who punished certain crimes. Their hair was made from serpents. If anyone asked where Hades was located, the answer had to be that it lay very deep below the ground.

This idea of an underworld is very widespread. In Hebrew, Sheol is the "abode of the dead," or underworld. The underworld is also the Christian hell and the *jahannam* of Islam and occurs in many other cultures. The dark fields of Norse mythology, the Niðavellir, are a land below the ground, inhabited by dwarfs. The Hindu and Buddhist *naraka* is located underground, but it is not a place of eternal punishment. It is a place of purification, like purgatory, not a hell.

At various times scientists have also subscribed to their own version of the idea. In 1692 the eminent English astronomer Edmund Halley (1656–1742) suggested that beneath the outer crust, which he believed to be about 500 miles (800 km) thick, the Earth consisted of two hollow shells, one inside the other and both enclosing a central core. He intended this as a solution to the problem of local variations in the Earth's magnetic field, which deflect a compass needle. In fact, it is the magnetic attraction of iron and nickel deposits that cause the variations, but Halley proposed that each of his shells had its own magnetic poles and the two shells rotated at different speeds. Each shell had its own atmosphere and its own internal source of light, and possibly the inner shells were inhabited. He thought that it was gas escaping from the shells that produced the northern and southern lights.

One of the most famous proponents of a hollow Earth was the American eccentric John Cleve Symmes (1779–1829). He arrived at his theory in 1818, holding that the Earth consisted of an outer shell, about 800 miles (1,287 km) thick, with four more shells inside it. All of the shells had openings about 1,400 miles (2,253 km) across at the North and South Poles. Symmes lectured all over the United States and even proposed expeditions to the poles to find the openings.

There were several other believers in a hollow Earth, with humans living on the outside of the spheres. Some believed the opposite, however, maintaining that we were living on the inner surface of a hollow sphere and that the stars and planets were contained within the sphere. An American doctor, Cyrus Reed Teed (1839–1908) propounded this idea and even founded a system of belief, called Koreshanity, on it (*Koresh* is the Hebrew equivalent of *Cyrus*).

Teed believed that gravity does not exist and that objects are held on the inner surface of the sphere by "centrifugal force" (in fact, inertia) due to the Earth's rotation. However, calculations show that this would be far too small to account for the gravity actually experienced, and it would decrease with distance from the equator. In fact, objects on the inside of a hollow Earth would be almost weightless if, as Teed thought, gravity did not exist. If gravity did exist it would be impossible to live on the inside surface of the sphere, because gravity would drag everything toward the center of the Earth.

THALES—AND THE FLOATING EARTH

The question of the composition and structure of the Earth is closely related to the nature of matter itself, and the ancient Greeks were the first people to give this serious consideration. Their philosophers held that all matter is composed of mixtures of four elements: earth, air, fire, and water. These were not the material substances the names suggested, but abstract concepts, often called "principles." To some extent these can change from one to another. Air can be compressed until it changes into water, for instance, and water can be changed into earth.

It all began with Thales, who lived in Miletus (*see* "Anaximander— and the First Map," pages 18–20). Little is known about him. He is most famous for having predicted a solar eclipse that occurred on May 28, 585 B.C.E. His prediction was not so impressive as it seemed. The Babylonians had discovered at least 200 years earlier that solar eclipses recur in a cycle of approximately 19 years, allowing them to predict a date when an eclipse might be expected, although they could not tell where it would be visible. Thales' knowledge probably went no further than this, either.

Thales believed that water was the most fundamental of all the elements and that the Earth was a disk floating on water. Thales was seeking an explanation for earthquakes. If the Earth floats on water, he speculated, then earthquakes might be due to large waves that rock the Earth. Herein lies Thales' importance. The reason he is often described as the first scientist is that he believed natural phenomena can be explained naturally. Causes reliably produce effects. This view, which is essentially modern, contrasts with the mythological explanations for phenomena prevailing at the time, such as that earthquakes and volcanic eruptions were the actions of such supernatural forces as gods or demons. Thales' saying "To bring surety brings ruin" enjoins us to cultivate an inquiring mind and beware of being totally certain about anything. That might be taken as a definition of the scientific attitude to the world.

RENÉ DESCARTES—AND THE WATERS UNDER THE EARTH

René Descartes was one of the most famous and influential philosophers and mathematicians of his generation. As a mathematician, he discovered that the location of any point on a two-dimensional surface can be specified by two numbers, now called the Cartesian coordinates. A grid reference and the use of longitude and latitude to specify location are applications of Cartesian coordinates. Descartes also found that if the value of one variable (x) depends on the value of another (y), so the relationship between them can be expressed algebraically (for example, $x = y^2 - 3$), the equation can be plotted on a graph, with changing values of x along the horizontal axis and the resulting values of y along the vertical axis. Descartes had merged algebra and geometry to create what came to be known as analytical geometry.

Descartes was born into a prosperous family on March 31, 1596, at La Haye, Touraine, in western France. His education began at a Jesuit college at La Flèche, which he entered when he was eight, and in 1612 he enrolled at the University of Poitiers to study law, graduating in 1616. He then joined the army of Prince Maurice of Nassau (then part of Germany) as a way of seeing something of the world and became a military engineer. He returned to France in 1622, sold the estate he owned at Poitiers, and spent the next few years traveling through Europe and meeting many scientists, finally settling in the Netherlands in 1629. In 1649 he moved again, this time to Stockholm, as Queen Christina

René Descartes, 17th-century philosopher and mathematician *(Sheila Terry/ Science Photo Library)*

of Sweden had asked him to instruct her in philosophy. Christina stipulated that her lessons commence every day at 5:00 A.M. This was a strain for Descartes, who seldom rose before noon and did much of his work in bed, and he suffered from a respiratory weakness. The early morning cold of the Swedish winter proved too much for him, and he died from pneumonia on February 11, 1650.

Although he was principally a mathematician and philosopher, Descartes had wider interests. He proposed a theory of the universe, explaining the formation of stars and planets. It was completely wrong, but it was accepted in France for more than a century. He also had ideas about the structure of the Earth and drew two cross-sectional diagrams showing the Earth consisting of five layers. According to a widely held view with which Descartes agreed, the Earth had once been molten, and the layers formed as the planet cooled and its rocks solidified.

Two features of Descartes's drawings are particularly interesting. The first is that one of them shows the subsurface layers buckled and broken. Descartes realized that huge forces must have acted on the rocks to produce the steeply inclined strata and faults that were evident in mountains such as the Alps. The second feature was that one of the layers consisted of water. Many people believed there were large quantities of water below ground. The widely accepted view was that all rivers begin as mountain streams fed by water breaking through the surface from a vast subterranean reservoir, like blood flowing from a wound. Descartes was one of the first to attempt to explain what was happening.

The existence of subterranean water was self-evident. People dug wells that filled with water, water could be seen rising from below ground at springs, miners often had to wade through water and mines sometimes flooded, in some parts of the world geysers shot water and steam high into the air, and volcanoes discharged steam when they erupted. But where did all this water come from, and how was it replenished? Descartes's theory was that seawater moved through underground channels and into large caverns below the tops of mountains. From there it emerged at the heads of mountain streams and springs. Descartes was quite wrong, but most scientists were not sufficiently interested even to attempt an explanation.

The English naturalist John Ray (1627–1705), a contemporary of Descartes, came closer to a correct understanding of underground water. Ray suggested that water descends from the surface through

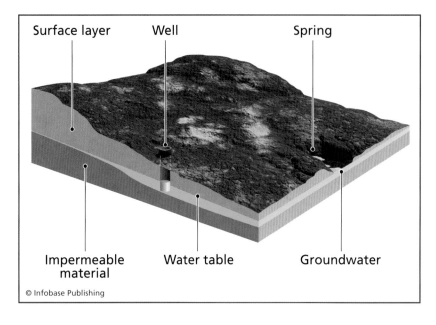

Surface layer   Well   Spring

Impermeable material   Water table   Groundwater

© Infobase Publishing

Water below ground, showing the groundwater and water table, and a spring

pores and small cavities then moves horizontally through gaps between rocks, the underground channels merging until eventually the water flows as a stream. Apart from Ray, however, it was not until the 19th century that a full explanation was found for the movement of underground water.

Ray was approximately correct. Rainwater drains vertically through the soil until it encounters a layer of impermeable material, where it collects as *groundwater*. Rock strata are rarely horizontal, and the groundwater lying on top of the impermeable layer flows slowly downhill through pores in the overlying, permeable layer. The groundwater saturates the permeable material through which it flows, and the upper surface of the saturated layer is called the *water table*. A hole dug from the ground surface that penetrates the water table will fill with water to the height of the water table. It is then a well. If the ground level slopes more steeply than the impermeable layer below ground, it may intersect the water table. Where that happens groundwater will break through the surface as a spring or emerge as a seep that saturates a small patch of ground. The diagram shows the arrangement of surface layer, impermeable layer, groundwater, and water table.

STRABO—AND HIS EXPLANATION OF VOLCANOES AND EARTHQUAKES

Thales believed that earthquakes happened when large waves rocked the Earth from below; earthquakes were manifestations of storms in the ocean on which the flat Earth rode. Many years later, Strabo (64 B.C.E.–21 C.E.) had a quite different idea. He believed that hot winds blew below ground, and from time to time these winds produced huge, powerful gusts that burst through the ground and into the air, carrying with them ash, dust, steam, and fire, at the same time causing the ground to tremble. Hot subterranean winds were the cause of volcanoes, and volcanoes shook the Earth, causing earthquakes.

Strabo's theory is less fanciful than it sounds, although it is quite wrong. Any educated person living in the Mediterranean region would have heard stories about volcanic eruptions and earthquakes and is very likely to have experienced them. It is a seismically active part of the world, where Africa is pushing northward into Eurasia, and the Mediterranean volcanoes are famous.

The very word *volcano* is from the name of the Roman god of fire, Vulcan, and the small island of Vulcano, off Sicily. Volcanoes erupt hot gases and steam, so it is not unreasonable to suppose there is hot air below ground that occasionally escapes in the form of huge gusts of wind, and volcanic eruptions are certainly linked to earthquakes.

It is unlikely that Strabo reached this conclusion by himself. He was a renowned geographer and historian, but he obtained much of his information from earlier writers, whose works are now lost. He claimed to have traveled from Armenia to Tuscany and from the Black Sea to the border of Ethiopia. He described fountains of naphtha that could be seen near the Euphrates River, the seasonal rise and fall of the Nile, and volcanic landscapes in Sicily and southern Italy, and he recognized that Vesuvius was a volcano and not simply an ordinary mountain, even though it had not erupted within living memory.

Strabo was born at Amaseia (modern Amasya) in what is now northern Turkey. His mother's family was very influential, and Strabo received a very good education, initially from Aristodemus, a former tutor of the sons of the military commander and civil administrator Pompey the Great (106–48 B.C.E.). In 44 B.C.E. Strabo moved to Rome to continue his studies under Tyrannion, who had formerly taught Cicero, and Athenodorus, a tutor of Octavius (later the emperor Augustus). Athenodorus probably introduced Strabo into the social circle surrounding the emperor. He remained in Rome until about 31 B.C.E., then left to begin his travels.

While he was in Rome, Strabo became a Stoic. Stoicism is a philosophical system teaching that all virtue is based on knowledge and defining knowledge as the harmonization of the individual's ideas with reality. A virtuous person, therefore, is one who lives in harmony with nature in full awareness of nature's guiding principle, which is reason. Reason is one of the attributes of God. All worldly concerns are illusory; nothing but virtue matters.

Strabo's major work, called *History*, was published in about 20 B.C.E. It came to 47 volumes, but only fragments survive. His 10-volume *Geography* survives in full, was probably completed about 7 B.C.E., and may have been an appendix to the *History*.

Greek philosophers could theorize about the nature and causes of earthquakes, but they had no means of detecting earthquakes that were too small to be felt. They lacked scientific instruments, and the first earthquake detector was made in the year 132 C.E., not in Greece, but in China (see facing sidebar).

MILNE, HIS SEISMOGRAPH, AND THE INTERIOR OF THE EARTH

The interior of the Earth is completely inaccessible. Miners dig into the ground to retrieve coal, and in some parts of the world they cut downward into hard rock in search of metallic ores, but the deepest mines penetrate only a short distance into the Earth's crust. The world's deepest mine is the East Rand gold mine in South Africa, where miners work 11,762 feet (3,585 m) below the surface. There are plans to extend the Western Deep mine, another South African gold mine, to 16,400 feet (5,000 m). Perhaps, in years to come, mining technology will advance sufficiently to allow miners to tunnel even deeper, but there are formidable problems to overcome. For one thing the temperature increases with depth, and at 16,400 feet (5,000 m) it reaches 158°F (70°C), so massive cooling equipment is needed. There is also the pressure caused by the immense weight of overlying rock. When rock is removed to make mine galleries, the change in pressure combined with the cooling can make the surrounding rock explode, in a phenomenon called *rock burst*. Rock bursts are responsible for many of the fatal accidents in deep mines cutting through hard rock.

It is possible to drill deeper boreholes, because these are narrower and no person has to descend through them. The world's deepest is the Kola Superdeep Borehole, on the Kola Peninsula, which projects into the White Sea off northern Russia. In 1989 that borehole reached a depth of 7.62 miles (12.3 km). That sounds impressive—and it is—but the Earth's radius is 3,959 miles (6,371 km), so the deepest

ZHANG HENG—AND HIS EARTHQUAKE WEATHERCOCK

Zhang Heng (also spelled Chang Heng; 78–139 C.E.) was a poet, painter, literary scholar, astronomer, mathematician, and inventor. A lunar crater and an asteroid bear his name, and even a mineral, zhanghengite, acknowledges his fame.

Zhang was born into an influential family in Nanyang, in Henan Province, in southwestern China. It is uncertain where he died. He was brought up in the political and moral tradition of Confucianism and left home at 16 to pursue his studies in the capital, which was then the city of Luoyang, in northern Henan. Zhang spent 10 years studying literature and training to be a writer, becoming a distinguished poet and author of more than 20 works that brought him fame.

At the age of 30 Zhang became interested in science and particularly in astronomy. In 116 C.E. he was appointed to an official position at the court of the emperor. Times were troubled. At the time of Zhang's birth the emperor was Zhangdi, an intelligent, benign emperor who ruled until the year 88. Zhangdi was succeeded by a series of nine emperors, the oldest of whom was only 14 when he acceded to the throne. Some were infants who survived for no more than one year. The emperors' maternal relatives and their retainers ran the court and the country.

Emperors received their authorization to rule from heaven, and to establish their link with the Earth one of their duties was to change the calendar. The task of devising a new official calendar fell to Zhang, and in 123 he succeeded in reforming the Chinese calendar, harmonizing it with his astronomical observations.

Zhang was also chief astrologer to the court. One of his responsibilities was to detect signs of bad government. The gods responded to bad government by causing earthquakes, and it was Zhang's job to detect them wherever in the empire they might occur. To help with this, in 132 he invented a device he called *houfeng didong yi* (instrument for measuring the seasonal winds and the movements of the Earth); in other words, it was an "earthquake weathercock" and the world's first *seismometer*.

In 2005 Chinese seismologists announced that they had constructed a replica of Zhang's original instrument. The illustration shows how it worked. At the center of an egg-shaped copper vessel there was an inverted pendulum with a weight at the top and a spike on top of the weight. The pendulum was carefully balanced, and if a tremor disturbed it, the weight fell toward one of eight channels that passed through the side of the

(continues)

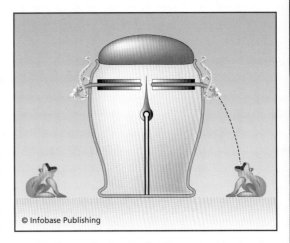

© Infobase Publishing

A reproduction of the earthquake weathercock invented by Zhang Heng in 132 C.E.

(continued)

vessel, and the spike entered the channel and moved along it. Each channel contained a slider, and at the outer end of each channel there was the figure of a dragon holding a copper ball in its mouth. The spike pushed the slider along the channel, and the slider dislodged the ball from the dragon's mouth. The ball fell into the mouth of one of the eight toads positioned around the base. When the ball hit the toad, it made a noise to attract attention. The sound of the falling ball indicated that an earthquake had occurred, and the particular toad holding the ball indicated the direction of the *epicenter*.

In February 138, Zhang's earthquake weathercock detected a quake that had not been felt in Luoyang, and he informed the emperor. He could even tell the emperor that the earthquake had occurred to the west of the city. He once reported an earthquake that had happened more than 600 miles (1,000 km) northwest of the city. News of that earthquake reached Luoyang several days later.

Zhang described his instrument in a work where he also described the universe as resembling a hen's egg, with the Earth lying alone at the center, like the yolk. "The sky is large," he wrote, "and the Earth is small."

hole in the world has penetrated about 2 percent of the distance to the center.

Nowadays geologists know a great deal about the structure of the Earth. They have measured the thickness of the crust, the depth of the *mantle* beneath the crust, and the dimensions of the outer and inner core, and they have achieved all this without ever visiting these regions. The instrument they use to study the interior of the Earth is called a seismometer, and seismometer readings are recorded by a *seismograph*.

Seismometers and seismographs detect and record *seismic waves*, which are the shock waves generated by earthquakes or large explosions that propagate through the Earth. Zhang Heng invented the first seismometer in 132 C.E. It was 1880 before a Western scientist improved on Zhang's invention. That scientist was John Milne.

Born in Liverpool, England, on December 30, 1850, Milne was educated at King's College, London, and the Royal School of Mines, and he worked as a geologist in Newfoundland and Labrador before, in 1874, taking part in an expedition to Egypt and northwestern Arabia.

In 1875 Milne was invited to become professor of geology and mining at the Imperial College of Engineering in Tokyo. He arrived in Japan in 1876, having traveled overland across Siberia by train and camel train partly because he enjoyed the adventure but also because he suffered from seasickness. He remained in Japan for 20 years. There was an earthquake on the evening he arrived in Japan, and in the succeeding years he had ample opportunity to study earthquakes, for Japan suffers more of them than any other country. Milne married a Japanese woman, Tona, and might have spent the rest of his life there had it not been for a fire that destroyed his home, laboratory, library, and many of his instruments on February 11, 1895. He returned to Britain with his wife and Japanese assistant, Mr. Hirota, and settled at Shide Hill House, near Newport, Isle of Wight, where he died on July 31, 1913. The illustration shows Professor Milne—nicknamed Earthquake Milne—and Tona, his wife, as they were in about 1900.

Milne was one of a small group of British geologists at the Imperial College. Following an earthquake in 1880 that caused some damage in Yokohama, Milne persuaded his colleagues to form the Seismological Society of Japan, which was the world's first society of *seismology.* He refused the presidency, however, maintaining that post should be held by a Japanese scientist. He edited the society's newsletter and wrote most of it.

In that same year, 1880, in collaboration with Alfred Ewing, a professor of mechanical engineering, and Thomas Gray, a professor of electrical engineering, Milne invented a seismograph. Essentially it consisted of a pendulum that was fixed at one end into solid bedrock. If an earthquake made the rock move, that movement was transferred either to a pen writing on a rotating cylindrical drum, or to a beam of light. In the years that followed Milne set up 968 seismic stations across Japan, all equipped with his seismographs, and conducted a seismic survey of the country. Following his return to Britain, Milne persuaded the Royal Society to finance the establishment of 20 earthquake observatories

John Milne and Tona Milne, his wife, ca. 1900 *(Science Museum/Science & Society Picture Library)*

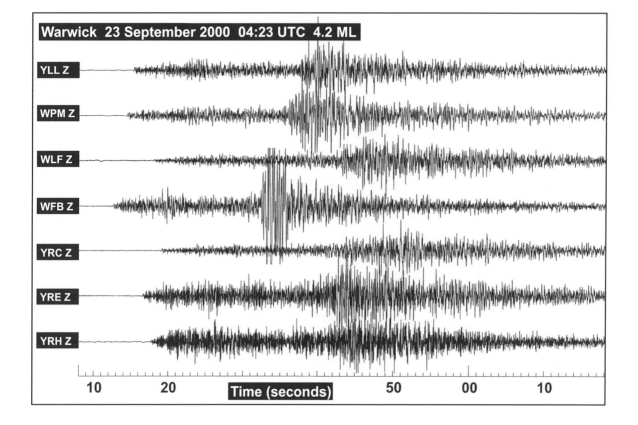

Warwick 23 September 2000 04:23 UTC 4.2 ML

YLL Z

WPM Z

WLF Z

WFB Z

YRC Z

YRE Z

YRH Z

10   20   **Time (seconds)**   50   00   10

Seismic trace recorded by the British Geological Survey of an earthquake that occurred near Warwick, England, on September 23, 2000 *(James King-Holmes/Science Photo Library)*

in different parts of the world at a cost of about $5,000. The network eventually expanded to 40 observatories, with their global headquarters in the Isle of Wight.

Milne had shown the way, and to this day seismographs record earthquake movements as ink traces on a rotating drum. The illustration shows the seismic trace recorded by the British Geological Survey of an earthquake with a magnitude of 4.2 on the Richter scale that occurred near Warwick, England, on September 23, 2000.

A seismograph records vibrations in the Earth. An earthquake happens when strains that have built up inside rocks finally overcome the strength of those rocks. The rocks break, and the bodies of rock on either side of the fracture move in relation to each other. That is an earthquake, and the place in the crust where it occurs is called the *hypocenter,* or focus. Rock movement at the hypocenter produces movement at the surface, directly above the hypocenter. The point where that happens is called the epicenter.

An earthquake generates waves, a little like the ripples that occur when someone throws a pebble into a pond but of several types and propagating spherically rather than across a two-dimensional surface. Rocks absorb some of the wave energy, and waves that are not absorbed pass through rocks of different types.

There are three principal types of seismic waves. P-waves (primary, or pressure, waves) are sound waves. They vibrate in the direction of wave travel and move at the speed of sound. S-waves (secondary, or shear, waves) vibrate in a direction perpendicular to the direction of travel and move at about 60 percent of the speed of P-waves. S-waves attenuate (weaken) more rapidly than P-waves. P- and S-waves are known as *body waves*. Surface waves travel across the surface away from the epicenter, moving at about 90 percent of the speed of S-waves, or 54 percent of the speed of P-waves. Seismologists are able to identify the types of wave producing seismograph traces, and by timing their arrival at the center from the seismometers detecting them, they can calculate the location of the hypocenter and epicenter.

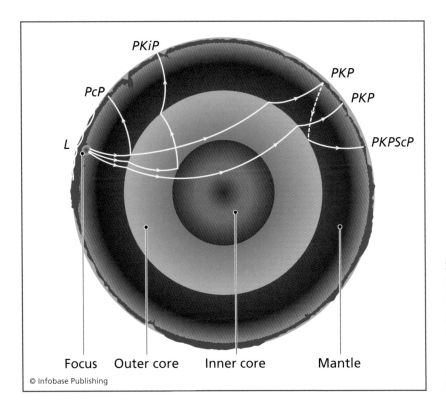

PKiP
PKP
PcP
PKP
L
PKPScP

Focus   Outer core   Inner core   Mantle

© Infobase Publishing

Seismic waves. Solid lines indicate P-waves; broken lines indicate S-waves. L = surface waves; c = waves reflected at the outer core; K = P-waves in the outer core; i = waves reflected at the inner core

As they encounter boundaries between the layers of the Earth, body waves can be reflected and refracted. Their speed also changes, because the speed of sound varies according to the density of the medium through which it travels—the denser the material, the faster sound travels. This alters the travel times of the waves, but it also provides seismologists with much more information about the materials through which the waves have passed. As the diagram shows, the arriving waves are then given additional labels. The full list of labels is shown in the table. Rayleigh and Love waves are surface waves with characteristics that were first described by the scientists whose names they carry (John William Strutt, Lord Rayleigh [1842–1919] and Augustus Edward Hough Love [1863–1940]). Love waves (also called Q-waves) make the ground surface move horizontally; Rayleigh waves (also called R-waves) make it move vertically, like surface waves on a pond.

S-waves cannot pass through fluids, but P-waves can. As S-waves pass through the outer core, they change into P-waves then change back into S-waves as they cross back over the boundary between the outer core and the mantle. This shows that the outer core is liquid. Studies of seismic waves have also revealed that the Earth's inner core is solid and that the mantle is made from very dense rock. They have also allowed seismologists to locate the boundaries between the crust and mantle, mantle and outer core, and outer core and inner core.

All of these discoveries derive originally from Earthquake Milne's invention of the seismograph. In 1906 Milne tried to measure the

SEISMIC WAVE CODE	
P, S	Within crust or mantle
p, s	Traveling upward from source
c	Reflected from outer core
K	P-wave that passed through outer core
i	Reflected from inner core
I, J	P- and S-waves that passed through inner core
LR	Rayleigh surface waves
LQ	Love surface waves

velocity of seismic waves through the deep layers of the Earth. He had only limited success with this. The person who found a way to abstract information from the seismic record was the Croatian geologist Andrija Mohorovičić (pronounced mo-ho-ROE-vich-ich).

MOHOROVIČIĆ—AND HIS DISCONTINUITY

On October 8, 1909, an earthquake struck Croatia, with its epicenter 24 miles (39 km) southeast of Zagreb. A network of seismometers had been installed earlier, and seismographs had recorded a complete set of data from the earthquake. Mohorovičić studied the data and made a number of discoveries from them.

First he found that earthquakes produce two types of waves, one that oscillates longitudinally and one that oscillates transversely. These are the waves now known respectively as P- and S-waves.

Mohorovičić also observed that seismic waves that penetrated deep into the Earth arrived at the seismometer ahead of those that had traveled along the surface, even when he had allowed for the different distances involved. This suggested to him that the outermost crust, through which the surface waves moved, rested above a layer of much denser material in which the waves moved more rapidly. He found that when seismic waves reach the boundary between materials of different densities, they behave like light rays and are reflected or refracted.

The behavior of the waves showed that the boundary between the upper crust and the denser material beneath it was abrupt. An abrupt change of this kind is known as a discontinuity, and this one is called the Mohorovicic discontinuity, or Moho for brevity (and simplicity of pronunciation). Mohorovičić calculated that the discontinuity lay 33 miles (54 km) below the surface; in fact, it is 15–37 miles (24.1–59.6 km) beneath the surface of continents and 3–6 miles (5–9 km) below the ocean floor. In 1961 the first of five holes was drilled off the coast of Guadalupe, Mexico, in an attempt to penetrate the upper crust to the discontinuity. The deepest reached only 601 feet (183 m), however, and in 1966 Project Mohole, as it was called, was abandoned.

Mohorovičić was a remarkable man. He was born on January 23, 1857, at Volosko near Opatija. His father was a blacksmith who specialized in making anchors, and Mohorovičić grew up with a love of the sea. His wife, Silvja Verni, was the daughter of a sea captain. The couple had four children.

Mohorovičić began his education in local schools then studied physics and mathematics at the University of Prague. He had a gift for languages, speaking Italian, French, and English by the time he was 15 and later learning German, Latin, and Greek.

After graduating he became a teacher, first at the gymnasium (high school) in Zagreb and then at a school in Osijek. In 1892 he took a post at the Royal Nautical School in Bakar. He remained there for nine years, and during that time he became interested in, and taught, meteorology. He studied the movements of clouds, which indicate the wind direction in air that is well clear of surface obstructions. In 1891, at his own request, he moved to the secondary school in Zagreb. In 1892 he became head of the Meteorological Observatory at Grič and established a meteorological service for the whole of Croatia. At the same time he taught geophysics and astronomy at the university. His observations of clouds provided the material for his doctoral thesis, and he was awarded his doctorate of philosophy from the University of Zagreb in 1893. He became a full member of the Croatian Academy of Sciences and Arts in 1898. Mohorovičić retired in 1921 and died at Zagreb on December 18, 1936. A lunar crater and an asteroid are named in his honor.

GOLDSCHMIDT—FOUNDER OF GEOCHEMISTRY

The rocks of the Earth's crust sustain life. Soil is made from mineral particles derived from rocks mixed with organic material contributed by plants and animals. Crustal rocks also supply industrial chemicals, such as superphosphate fertilizer obtained from calcium phosphate rocks and gypsum (hydrated calcium sulfate, $CaSO_4.2H_2O$) used in the production of Portland cement and plaster of Paris. Alabaster, used by sculptors, is a pure white variety of gypsum. Rocks also supply the metals used to make tools and weapons. A few metals occur in a pure form, as native elements, but even these usually have to be pried from the rocks in which they are embedded. Most metals are obtained from ores, by heating a rock to melt the metal it contains or by treating it chemically.

The solid Earth is the source of all the resources without which a modern society could not function. It supplies our food, grown in soil. Uranium, a metal derived from rocks such as uraninite, formerly known as pitchblende (UO_2), is used in nuclear reactors to produce heat to drive turbines that generate electricity. Its rocks even provide tokens of wealth in the form of gemstones and the metals to make coins.

Mineral resources are not distributed evenly around the world, so they are traded between nations. In times of war combatant nations often blockade one another's ports to starve their opponents of essential supplies of these resources, creating scarcities. During World War I (1914–18) metals and minerals became scarce, and this attracted the attention of Victor Moritz Goldschmidt, a young Norwegian chemist. He asked some very fundamental questions. What are rocks and minerals made of? How do they form? Is it possible to predict which rocks are likely to contain useful amounts of valuable substances? Goldschmidt studied the chemistry of rocks and minerals—the chemistry of the Earth. He is widely regarded as the founder of the scientific discipline of Earth chemistry, or *geochemistry.*

Goldschmidt was born on January 27, 1888 in Zurich, Switzerland. His father, Heinrich, was a physical chemist, and in 1901 he was appointed professor of physical chemistry at the University of Kristiania (now the University of Oslo). The family moved to Kristiania (as Oslo was then known). At that time Norway and Sweden were united. Norway became independent in 1905, and the Goldschmidts became Norwegian citizens. Goldschmidt completed his schooling in Kristiania then studied geology at the University of Kristiania. He was a remarkable student. In 1909, aged 21, he obtained a postdoctoral position without taking the degree examination, and in 1911 he was awarded his doctorate, which is a qualification most people did not obtain until they were in their 30s or 40s. In 1873 the Dutch physicist Joannes Diderik Van der Waals (1837–1923) had discovered the *phase* rule governing the relationship between the number of phases, number of components, and the number of *degrees of freedom* in a system that is in equilibrium. The phase rule applied to gases, but in his doctoral thesis Goldschmidt applied it to changes that take place in minerals in crustal rocks undergoing contact metamorphism—the melting and recrystallization that occurs in rocks heated by contact with magma. Goldschmidt became an associate professor of mineralogy and petrology in 1912, and in 1914 he was appointed professor of mineralogy and director of the Mineralogical Institute.

War broke out in 1914, and Goldschmidt and his colleagues embarked on a serious study of the distribution of elements in crustal rocks. During the years that followed they studied 200 compounds of 75 elements and produced the first table of the radii of atoms and *ions*. Goldschmidt showed that the size of its constituent ions is the most important factor determining the structure of crystals and applied this to explain the structure of complex minerals. Eventually he was able to predict the types of minerals in which particular elements should occur. With that discovery, mineralogy ceased to be a purely descriptive science. In 1929 Goldschmidt took up a post as professor of mineralogy at the University of Göttingen, Germany, where he continued his research, expanding it to estimates of the abundances of elements in the universe and the relationship between the stability of *isotopes* and the abundance of elements. The science of geochemistry was thus firmly established.

Goldschmidt was Jewish, and in 1935, following the Nazi rise to power in 1933, he returned to Oslo (the name was changed from Kristiania in 1924) in order to protest the increasingly harsh restrictions the Nazis were placing on Jews in Germany. Arriving with no job, he worked in industry until 1936, when he found an academic position and was able to resume his research. Germany invaded and occupied Norway in 1942, however, and Goldschmidt was arrested. He was due to be deported to a concentration camp but escaped with the help of his colleagues and the Norwegian resistance. He went first to Sweden and then on March 3, 1943, he was flown to Britain by the Secret Intelligence Service. He passed on information about technical developments in occupied Norway. In August he was given a position at the Macaulay Institute for Agricultural Research in Aberdeen and in 1944 moved to Rothamsted Experimental Station in Hertfordshire, where he worked on soils. Goldschmidt returned to Oslo on June 26, 1946 and died there on March 20, 1947.

WALTER ELSASSER—AND THE DYNAMO IN THE CORE

Sailors navigate by remaining within sight of the coast or by observing the currents, depth, and appearance of the sea. They use the Sun, which at noon is always due south or north depending on whether the vessel is in the Northern or Southern Hemisphere, and at night they can steer by the stars. These methods have been used for thousands of years and they are very reliable—provided the sky is clear. If the sky is overcast and, even worse, if there is fog, until the introduction of satellite positioning the navigator had to rely on dead reckoning. This method involves monitoring the ship's speed and direction and using this information to plot its present position from its last known position. It works provided it is possible to steer the ship in a desired direction, but that is impossible without some external reference to use for guidance.

The Book of the Devil Valley Master, written in China in the fourth century B.C.E., mentions the fact that lodestone (now known as magnetite, an iron oxide) attracts iron. By the 11th century there are records of the Chinese use of a piece of magnetized metal to find the directions of north and south, and by early in the 12th century Chinese sailors were using a magnetic compass as a navigational tool. European sailors began using magnetic compasses in about 1300.

Compasses have been in use for so long and they are so familiar that it is easy to forget that until fairly recently no one had any idea how they worked. They just did. Rub an iron needle on a piece of silk, suspend it on a thread so it is horizontal, and it will point north and south. Fix a magnet to the underside of a disk showing north, east, south, and west, float the disk on a liquid, and the result is a mariner's compass. The illustration shows a typical design, in this case from a compass made in the 18th century.

A French scholar, Peter Peregrinus of Maricourt (*Peregrinus* means "pilgrim"; in Latin his French name was Pierre Pèlerin

Magnetic compass. A mariner's compass card made in the 18th century

© Infobase Publishing

de Maricourt), who flourished in about 1270 (his birth and death dates are unknown), was the first person to note that like magnetic poles repel each other and unlike poles attract. In *Epistola Petri Peregrini de Maricourt ad Sygerum de Foucaucourt, militem, de magnete* (Letter about the magnet of Peter Peregrinus of Maricourt to Sygerus of Foucaucourt, soldier), written on August 8, 1269, he explained how to identify the north and south poles of a magnet and also his discovery that it is impossible to isolate a magnetic pole by breaking a magnet in two; this simply produces two smaller magnets, each with a north and south pole. It was Peregrinus who invented the forerunner of the mariner's compass, but he made one mistake: Peregrinus believed that a compass points to the pole of the celestial sphere, a fixed point in space. Centuries later an English scientist, William Gilbert (1544–1603), discovered where the needle truly points.

Gilbert was a physician. He became president of the College of Physicians, and in 1601 he was appointed court physician to Queen Elizabeth I. Medicine is how he earned his living, but he had another interest: magnetism. In 1600 he published a book on the subject, *De Magnete* (About magnets). Gilbert was an experimenter. He smeared a magnet with garlic and found its properties were unaltered, thereby disproving the popular belief of the time that garlic destroys magnetism. Everyone knew that a magnetized needle points north and south, but Gilbert found that if it is allowed to move vertically, the needle also points downward, a deflection now known as *magnetic dip*.

Gilbert went further. He obtained a spherical magnet and found that when a suspended compass needle was carried across its surface, the angle of magnetic dip changed from 90° (pointing directly downward) over each of the sphere's poles, to 0° over the sphere's equator. This led him to suggest that the Earth itself is a spherical magnet and that the location of its poles is fixed. His work was so important that the unit of magnetomotive force is called the gilbert (abbreviated Gb). The Earth's magnetic poles are not in fixed locations, however. In 1635 the English astronomer and mathematician Henry Gellibrand (1597–1636), professor of astronomy at Gresham College, London, published his observation that the direction of a compass needle in London had shifted by more than seven degrees during a period of 50 years. The Earth's magnetic field moves; it also strengthens and weakens slowly, and from time to time its polarity reverses.

The Earth has a magnetic field, but how is it produced? That remained a mystery until the middle of the 20th century. Geologists knew that the Earth has a core made of iron, so at first it was assumed that the core was solid and a permanent spherical magnet, just as Gilbert had suggested. But then it was discovered that the outer core in fact is liquid, and although the iron inner core is solid, it is above the Curie temperature, named for Pierre Curie (1859–1906), who discovered it. As the temperature of iron rises, its atoms vibrate increasingly vigorously until eventually the coupling between them breaks. It is this coupling that gives iron its magnetic properties. Consequently there is a temperature—the Curie temperature, or Curie point—beyond which magnetism is destroyed. The Curie temperature for magnetite is 1,247°F (675°C). The Earth's core temperature is 9,000–12,600°F (5,000–7,000°C). Clearly the Earth's core is no ordinary magnet.

Walter Maurice Elsasser (1904–91) solved the problem in 1939 by suggesting that the Earth's core is not a permanent magnet but a dynamo. The inner core is solid and made from nickel and iron. It was probably magnetized by the Sun's magnetic field at the time it formed. Surrounding it the outer core is liquid, and because of the very high temperature, it flows very freely. Heat generated by radioactive *decay* produces convection currents in the outer core, and combined with motions produced by the Earth's rotation, these produce electric currents that sustain and amplify the magnetic field. The electric currents interact with the concentric shells of the Earth's mantle, which rotate at different speeds, and the *Coriolis effect* causes eddies in the currents at the boundaries between the shells. Details of the way currents in the fluid outer core flow are extremely complex and must account for the fact that the intensity and orientation of the geomagnetic field change with time. Periodically the field weakens, disappears altogether, then reappears with its polarity reversed—the Earth's Magnetic North and South Poles change places. Research continues, but it was Elsasser who pointed the way.

Elsasser was born at Mannheim, Germany, on March 20, 1904, and studied at the University of Göttingen, obtaining his doctorate in 1927. He accepted a teaching post at the University of Frankfurt but left Germany in 1933, following Adolf Hitler's rise to power. Elsasser went first to Paris, where he spent three years working on the theory of atomic nuclei at the Sorbonne. In 1936 he moved to the United

States to work at the California Institute of Technology. He became a U.S. citizen in 1940. During World War II (1939–45) Elsasser worked on radar for the U.S. Army Signals Corps and the Radio Corporation of America, returning to academic life when the war ended. He was appointed professor of physics at the University of Pennsylvania in 1947, head of the physics department of the University of New Mexico in 1960, professor of geophysics at Princeton University in 1962, and from 1948 until 1974 he held a research professorship at the University of Maryland. He also held positions at the University of Utah, Johns Hopkins University, and the Scripps Institution of Oceanography. He died in Baltimore, Maryland, on October 14, 1991.

Native Metals
and Metal Ores

People began using metals many thousands of years ago. At first it was because the metals were attractive and rare. Displaying metal adornments advertised social status and political power. Kings wear golden crowns, after all, not wooden ones. In time, however, people found other uses for metals. Particular metals could be shaped, made hard, and sharpened to a point or cutting edge. That made them useful as tools and weapons that were superior to those made from flint, antlers, bones, and wood. Once our ancestors began using metals—first copper, then bronze, then iron—their dependence on them increased. Modern civilization could not survive, at least in anything remotely resembling its present form, if metals could no longer be used. The table lists most of the metals, with the exception of some radioactive metals, many of which survive for only a very short time, by their year of discovery. It shows that the most basic metals were known in ancient times—it is impossible to assign a date to their discovery—and that many were discovered during the 18th and 19th centuries.

Gold was the first metal to find a use because it was the easiest to obtain, although it was found only in certain places. The other metals were discovered over the succeeding centuries as the technologies were developed for identifying, extracting, and refining them. This chapter explores the ways in which the search for metals has contributed to the understanding of the solid Earth.

YEAR OF DISCOVERY OF PRINCIPAL METALS

YEAR	METAL	CHEMICAL SYMBOL
*	Antimony	Sb
*	Bismuth	Bi
*	Calcium	Ca
*	Copper	Cu
*	Gold	Au
*	Iron	Fe
*	Lead	Pb
*	Mercury	Hg
*	Silver	Ag
*	Tin	Sn
1669	Phosphorus	P
1735	Platinum	Pt
1737	Cobalt	Co
1746	Zinc	Zn
1751	Nickel	Ni
1774	Manganese	Mn
1778	Molybdenum	Mo
1782	Tellurium	Te
1783	Tungsten	W
1789	Uranium	U
	Zirconium	Zr
1790	Strontium	Sr
1791	Titanium	Ti
1794	Yttrium	Y
1797	Chromium	Cr
1798	Beryllium	Be
1801	Niobium	Nb
1802	Tantalum	Ta
1803	Cerium	Ce
	Palladium	Pd
	Rhodium	Rh

YEAR	METAL	CHEMICAL SYMBOL
1804	Iridium	Ir
	Osmium	Os
1807	Potassium	K
	Sodium	Na
1808	Barium	Ba
	Magnesium	Mg
1817	Cadmium	Cd
	Lithium	Li
	Selenium	Se
1825	Aluminum	Al
1830	Vanadium	V
1844	Ruthenium	Ru
1860	Caesium	Cs
1861	Rubidium	Rb
	Thallium	Tl
1863	Indium	In
1875	Gallium	Ga
1879	Scandium	Sc
1886	Germanium	Ge
1898	Polonium	Po
	Radium	Ra
1923	Hafnium	Hf
1925	Rhenium	Re
1937	Technetium	Tc
1939	Francium	Fr
1940	Plutonium	Pu

* Known in ancient times

Metals exist in three forms. A pure metal is a chemical element, and approximately three-quarters of all the elements are metals. Pure metals can be mixed to make *alloys.* Brass, for example, consists of about 67 percent copper and 33 percent zinc, copper and zinc being elements (chemical symbols Cu and Zn, respectively). Metals also exist as compounds with other elements, and this is the form in which all but a very few occur naturally.

But what exactly is a metal? It is an element that is shiny and a good conductor of heat and electricity. The atoms of most metals lack their full complement of electrons, so they carry a positive charge and are said to be *electropositive.* This means that metals seldom form compounds with other metals but react readily with nonmetallic elements carrying a negative charge. A few metals do not react readily with other elements. Members of this group, including copper, gold, platinum, and silver, are sometimes found in their pure state. The most abundant metals in crustal rocks are, in descending order of abundance, aluminum, iron, calcium, sodium, potassium, and magnesium.

GOLD, WEALTH, AND POWER

Over the long course of history countless kings have ruled, died, and been forgotten. Only a few achieve immortality and some for curious reasons. Croesus (ca. 595–ca. 546 B.C.E.) is remembered for his fabulous wealth. Very wealthy people are still described as being "as rich as Croesus," and his name was already a synonym for a person of great wealth among the Greeks and Persians some 2,000 years ago.

Croesus was the last ruler of Lydia, a kingdom in what is now Turkey, and he was indeed extremely rich. His armies conquered Ionia, he formed an alliance with Sparta, and when the power of the Persian Empire increased under Cyrus the Great (reigned from 559 to 530 B.C.E.), Croesus joined forces with the Babylonians to oppose it. As well as being wealthy, Croesus was the powerful ruler of a large empire. His wealth derived from the large amounts of gold mined in Lydia and in the sands of the river Pactolus. The Lydians under Croesus are thought to have been the first people in the world to mint metal coins.

Croesus was a client of the Oracle at Delphi, and there are records of the rich treasures he left there. He consulted the Oracle when he was considering whether to attack Cyrus, and the Oracle gave him one of her most famous pieces of advice: "If Croesus crosses the Halys, a great empire shall be brought down." Croesus's army had to cross the Halys River to reach Persia, and legend has it that Thales of Miletus (see "Anaximander and the First Map," pages 18–20) altered

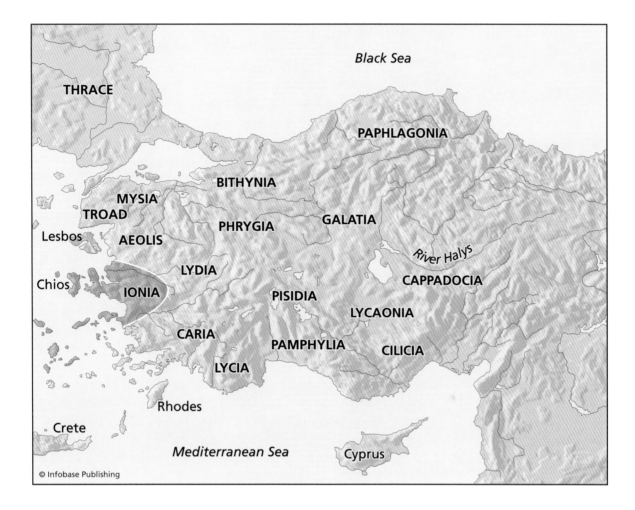

Black Sea

THRACE

PAPHLAGONIA

BITHYNIA

MYSIA
TROAD

PHRYGIA

GALATIA

Lesbos

AEOLIS

River Halys

LYDIA

Chios

IONIA

PISIDIA

CAPPADOCIA

LYCAONIA

CARIA

PAMPHYLIA

CILICIA

LYCIA

Rhodes

Crete

Mediterranean Sea

Cyprus

© Infobase Publishing

the course of the river to make the crossing easier. Encouraged by the Oracle, Croesus attacked, but when winter approached, his armies returned home, as was the custom at the time. Unfortunately for Croesus, the Persian army ignored the custom. With the Lydian army stood down, Cyrus attacked and defeated Croesus. A great empire was indeed brought down—the empire of Croesus, which was promptly absorbed into the Persian Empire.

Gold confers political power, but it also encourages foolishness, and Midas is the most famous example of a foolish king. He was ruler of Phrygia, a kingdom to the northeast of Lydia, but there were several Phrygian kings called Midas, and it is possible that it was a dynastic name, alternating with Gordieus. The Midas of legend lived

The ancient Near East, showing the location of Lydia, ruled by King Croesus. Ionia, shown in pink, was inhabited by the descendants of Greek colonists, so culturally it was linked to Greece.

in the eighth century B.C.E. The story goes that Silenus, the teacher and foster father of the god Dionysus, wandered off while drunk, and either a group of peasants discovered him and carried him to King Midas or he found his way into Midas' garden and passed out there. Midas recognized him and treated him as an honored guest for 10 days and nights before taking him back to Dionysus. Dionysus rewarded the king by offering him whatever he wished, and Midas asked that anything he touched should turn into gold. Dionysus granted the wish, and the delighted Midas touched a tree, turning it to gold. On returning home he ordered his servants to set a great feast on the table. They did so, but when he touched the food and drink, it turned to gold, and he faced starvation. In one version of the story he touched his daughter, and she turned into a golden statue. Midas cursed his gift and prayed to Dionysus to free him from it. Dionysus told him he must bathe in the river Pactolus. He did so, and the river sands turned to gold; Midas was free, and thereafter the Pactolus was rich in gold.

There are certain parts of the world where river sands contain grains or nuggets of pure gold. In ancient times there were rich alluvial deposits in Lydia, Persia, India, China, and the lands around the Aegean. The earliest known source of gold was on the Black Sea coast near the modern city of Varna, in Bulgaria, and the metal was certainly being used in Egypt and Mesopotamia by about 4000 B.C.E. By 3000 B.C.E. rich people were using gold rings to pay their bills. Gold was the first metal to be exploited.

There has never been enough gold to satisfy the need or greed of the powerful, so anyone who discovers a source will certainly become wealthy. The 1848 discovery of alluvial deposits in California triggered a gold rush that continued until 1855 and brought approximately 300,000 people in search of the elusive riches. This was a major historic event, but it was far from unique. There have always been adventurers willing to risk all in the search for gold. Stories have been told of their exploits for thousands of years. Jason and his recovery of the Golden Fleece was one of the earliest (see facing sidebar).

Jason's story may record a real event, albeit one that storytellers embellished until it acquired the status of a myth. The *Argo* was a very large vessel, and perhaps the story tells of the first important and somewhat piratical maritime expedition, launched in the quest for rich spoils. Others have suggested that the fleece was genuine and was used to filter gold grains from a river somewhere in Asia Minor. The grains adhering to the wool would have transformed it into a truly golden fleece. In 2004 German and Georgian archaeologists working at Sakdrissi, about 44 miles (70 km) southwest of the Georgian capital Tbilisi, discovered what may be the world's oldest gold mine, dated to about 3000 B.C.E. It is possible that the gold, used

THE GOLDEN FLEECE

Long ago in the land of Thessaly, a local king, Athamas, grew tired of Nephele, his queen, and took another wife. Nephele, fearing for the safety of her young son and daughter, sought the help of Hermes, the messenger of the gods, who gave her a ram with a golden fleece. Nephele set her children on the ram's back, and at once it leaped into the air and flew away to the east. As the ram crossed the sea separating Europe from Asia, the girl, Helle, fell from its back into the sea, which from that time was known as the Hellespont (now the Dardanelles). The ram continued on its journey with the boy, Phryxus, finally landing in the kingdom of Colchis, on the shores of the Black Sea and well out of his stepmother's reach, where Phryxus was made welcome by the king, Aeëtes. Phryxus sacrificed the ram and presented its golden fleece to Aeëtes, who placed it in a sacred grove, watched over by a dragon that never slept.

Some time later Aeson, the king of another part of Thessaly, grew tired of governing his land and passed the crown to his brother, Pelias, on condition that he hold it only until Aeson's son Jason was old enough to become king. When Jason grew up and demanded the crown, Pelias pretended to be willing to part with it but suggested that Jason should first go in quest of the Golden Fleece, which was known to be in Colchis and which Pelias maintained really belonged to their family. Jason planned an expedition. He employed Argus to build him a ship capable of carrying 50 men—a huge vessel in those days—and named it the *Argo* in his honor. Jason invited all the adventurous young men of Greece to join him, and from them he recruited a crew of heroes, each with a special talent. The crew called themselves the Argonauts.

After many adventures the Argonauts reached Colchis, and Aeëtes agreed to give Jason the fleece, but first he had to yoke two fire-breathing bulls with brass feet, use them to plow a field, and sow in the furrows the teeth of a dragon that had been killed earlier. Helped by the king's daughter Medea, who was a powerful sorceress and whom Jason had promised to marry, Jason calmed the bulls, sowed the teeth, and when a hostile armed warrior sprang up from each tooth, he turned them to fight against each other. Finally, he put the dragon to sleep with a potion supplied by Medea, seized the Golden Fleece, and, with Medea at his side and accompanied by his crew, fled back to the *Argo* and set sail before Aeëtes could stop them. They reached Thessaly, and Jason presented the fleece to Pelias. What happened to the fleece after that, no one knows.

to make necklaces, rings, and needles, was panned from the nearby river rather than having been dug from the ground. To this day villagers in certain parts of Georgia pan for gold using sheep's fleeces. Was Georgia Jason's destination?

Gold is too soft to make tools that can compete with those made from flint and bone. Its value lies in its beauty and its incorruptibility—it does not tarnish. That is because the compounds that gold forms with oxygen are highly unstable, breaking down rapidly to release the oxygen and leave the gold unchanged. Consequently gold tends to occur naturally as the pure metal, and where it is mixed with other metals, it can be separated from them by oxidizing them. Its incorruptibility means that gold is virtually indestructible. At the same time it is soft and malleable, so it is easy to work and small amounts go a long way.

Its beauty and purity have given gold a religious significance, and it acquired many ritual uses, some of them extraordinary. Not

THE DREAM OF EL DORADO

In Spanish, *el dorado* means "the golden one," and in the 16th century many people believed there was a real person to whom this description could be applied, the ruler of a land so wealthy that his courtiers regularly threw gold and priceless jewels into a lake. El Dorado came to refer to a place, but originally it referred to a person known as "the golden Indian" (*el indio dorado*) or "the golden king" (*el rey dorado*), who lived in what is now Colombia. The strange thing about this myth is that it was based in fact.

In the northern Andes there were at that time two confederations of tribes, together called the Muisca. Each tribe retained its own identity. There was no overall ruler, but in times of trouble the warriors of the northern federation were under the control of a leader called the *zaque* and those of the southern federation, centered around Bacatá (modern Bogotá), were controlled by the *zipa*. The Muisca worshipped several gods, one of which lived in Lake Guatavita, a circular lake in what may be a meteor crater. The Muisca honored the lake with a ritual that may have been held to mark the installation of a new *zipa*. The *zipa* was carried on a raft to the middle of the lake together with several tribal chiefs. All were naked, and the *zipa* was covered with a sticky mud and then with gold dust. The ceremony involved casting gold and jewels into the lake, and the *zipa* dived into the lake, washing off the gold, after which the raft returned to shore and there was singing, dancing, and feasting. This ceremony was well known throughout northern South America.

In 1535 the Spanish explorer and conquistador Gonzalo Jiménez de Quesada (1509–79)

long after Spanish explorers and conquistadores arrived in Central and South America in the early 16th century they began to hear tales of a king covered in gold whose courtiers and priests threw gold and precious stones into a lake. Not surprisingly, they set of in search of this golden man, called in Spanish El Dorado (see sidebar).

The Europeans failed to find El Dorado, but other American societies used gold, and the invaders robbed palaces, temples, and graves to obtain it. Between the arrival of Christopher Columbus in 1492 and 1600, more than 247,500 tons (225 tonnes[t]) of looted American gold was shipped back to Europe.

COPPER AND TIN MAKE BRONZE

Few metals occur in their pure form, but copper is one of them. Gold is attractive but too soft to be of use for anything other than orna-

---

arrived in Colombia as chief justice to the Santa Marta colony. He led a party into the interior of the country, where he encountered the Muisca. Stories of the Muisca lake ceremony spread through the Spanish community, and expeditions set off in search of the fabled gold. As time passed, the story grew, and El Dorado became a country with golden buildings and two legendary cities, Manoa and Omagua. One Spanish traveler even claimed he had been shipwrecked and rescued by people who took him to Omagua where he was entertained by El Dorado himself.

There were many attempts to find El Dorado. A party led by Gonzalo Pizarro (1502–48) crossed the Andes from Quito in 1539; Francisco de Orellana (ca. 1500–ca. 1549) sailed down the Napo and Amazon Rivers in 1541–42, traveling the full length of the Amazon; and from 1569 to 1572 Jiménez de Quesada explored eastward from Bogotá. In 1595 Sir Walter Raleigh (1552 or 1554–1618) searched for Manoa in the Orinoco Basin. Local tribes were responsible for some of the embellishment to the story. They used it to get rid of Europeans by sending them in search of El Dorado.

No one ever found the land of gold, but the Muisca were wealthy by European standards. Colombia is still the world's leading producer of emeralds, and the region produces copper, coal, and salt. Oddly there is no gold in the territory occupied by the Muisca; they imported it, and in large enough amounts to allow them to use it in many of their handicrafts. It was objects made by their craftspeople that they offered to the goddess of Guatavita in the *zipa* ceremony. The Europeans did find the lake. The notch is still there that they carved in 1580 in an attempt to drain the water from it.

ments and tableware. Copper is harder. It can be fashioned into tools and weapons.

Copper was first used in this way, as a preferred alternative to stone implements, about 10,000 years ago. Metalworkers used stone hammers to beat native copper into shape, and when the metal was made into blades, these were sharpened on stones. By about 6,000 years ago Egyptian workers were melting native copper and casting it in molds.

The technology for melting and molding metals had already been established. Gold is soft enough to be worked cold, and silver was probably the first metal to be processed in this way. Silver occurs as the native element, and it melts at 1,764°F (962°C), which is a temperature that can be reached in a well-ventilated, open wood fire. Melting gold was a little more difficult because it melts at 1,947°F (1,064°C), and the temperature in an open wood fire reaches only about 900°F (480°C).

Wood was not the only fuel available, however. People had discovered that if they piled wood in a large heap, covered the heap with mud to exclude air, and set the wood on fire but made sure the fire smoldered without bursting into flames, they would change the wood into charcoal. The process heated the wood sufficiently to drive off moisture—all wood contains water, even if it feels completely dry—and other volatile substances, leaving behind an impure form of carbon. Charcoal burns at a much higher temperature than wood and a charcoal kiln that could melt gold could also melt copper.

Before long, supplies of native copper became scarce as the most accessible deposits were depleted, but by then early metallurgists had found an alternative source of the metal. Chalcopyrite, or copper pyrites, is a copper-iron sulfide ($CuFeS_2$). It is the color of brass and harder than gold, but anyone mistaking it for gold and melting it would produce not gold but copper. Chalcopyrite is one of the most abundant copper ores, but others include chalcocite, or copper glance (CuS_2); azurite ($CU_3(CO_3)_2(OH)_2$); bornite, or peacock ore (Cu_3FeS_4); and malachite ($Cu_2CO_3(OH)_2$). Apart from chalcocite, which is slate-gray or black, copper ores are brightly colored. Azurite is deep blue, bornite is purple or blue, and malachite is bright green. They are easy to recognize, and copper smelting (heating the ore to extract the metal) began at about the same time native copper was first being molded. Its chemical symbol, Cu, stands for *cuprum,* the Latin name for copper; *cuprum* is a corruption of the earlier name *aes Cyprium,* which means "metal of Cyprus," referring to the source of most of the copper used in Roman times.

Copper is an attractive metal, but tools made from it are not very practical. Sharp edges soon become dull, and even modest pressure will bend blades and tines, but this problem was solved more than 5,000 years ago with the invention of *bronze.* Bronze is an alloy of copper and tin. The proportions vary, but early bronze was about 67–95 percent copper. Bronze church and cathedral bells use 75–80 percent copper.

Bronze figure of a cat and her kitten from ancient Egypt; the figure is held at the Louvre Museum, Paris, France. *(Erich Lessing/Art Resource, NY)*

Tin melts at 450°F (232°C), so smelting it was not difficult. The difficulty was the scarcity of tin. The main sources in the ancient world were in Cornwall, in the southwest of Britain, and Spain. Bronze was a great improvement on pure copper; not only is bronze harder than copper but also it is easier to melt for casting, a property that makes it especially suitable for making statues or special figures. The photograph is of a charming bronze figure of a cat and her kitten that was made in ancient Egypt. Bronze is also harder than iron and much more resistant to corrosion. In time bronze acquires a surface sheen, but it does not corrode away entirely, as iron does. Bronze is also used to make coins. "Copper" coins in fact are made from bronze, with about 95 percent copper, 4 percent tin, and 1 percent zinc.

Bronze is superior to iron, although not to steel, and the Bronze Age, during which bronze was the most important metal, gave way to the Iron Age not because of the inherent superiority of iron but simply because iron is more abundant than either copper or tin. Its abundance made iron cheaper, and before long it was more widely available.

IRON AND THE NEED FOR FUEL

As people in many different parts of the world learned to work with metal, to extract metals from their *ores,* and to alloy metals, they were also gaining a deeper understanding of the world around them.

In order to smelt ores it is first necessary to identify those ores, to distinguish one type of rock from another by its appearance, and to know how to build a fire that will generate sufficient heat. Metalworkers had to appreciate the properties of the metals they were using, and the intricacy and beauty of the metal objects, thousands of years old, displayed in museums, attest to their skill. Eventually, however, metal workers faced a challenge: Gold, silver, and copper are soft and easy to fashion into articles; bronze is hard but easy to cast in molds; iron is different.

Iron melts at 2,795°F (1,535°C), a temperature unattainable with a furnace of the type suitable for melting copper, and impurities mixed with the metal alter its physical properties. Hammering a cold piece of copper into shape hardens the metal, but the processing of iron is more complicated. Iron is used in two forms: *wrought iron* and *cast iron*. Cast iron is made by pouring molten iron into a mold of the desired shape. Wrought iron can be hammered, stretched, and twisted into a desired shape, and prolonged hammering renders it hard and brittle. It can then be returned to its more malleable state by heating it to red heat and allowing it to cool slowly, a process known as *annealing*.

Native iron is extremely uncommon; it occurs in a few rocks but also in a class of meteorites, where it is often mixed with 4–30 percent nickel. The earliest known ironwork consists of beads found at Jirzah, Egypt, which were made about 5,500 years ago from meteoritic iron. Meteorites reach the Earth from space, and they can fall anywhere. Most meteorites are made from rock, however. Iron meteorites are rare, and the metal would have been valuable and perhaps magical, given its origin. Many meteorites are recovered by people who watch them fall from the sky. The metal became much more common with the discovery of iron ores and the techniques for *smelting* them. The Hittites, living in Mesopotamia (modern Iraq), were producing iron by smelting ores by about 4,000 years ago. From Mesopotamia the technology spread to Greece, reaching the lands around the Aegean Sea about 3,000 years ago and the remainder of Europe about 2,600 years ago. One of the pieces of iron from that time is a dagger with an iron blade and bronze hilt.

For thousands of years painters have used hematite as a pigment. Hematite is the most widely distributed iron ore, and if a fire was lit beside a hematite-containing "paint rock," a party of artists or paint makers might have inadvertently smelted a small amount of the metal from it. When the fire burned out, they would have found small lumps of shiny metal among the ashes. By the beginning of the Bronze Age, around 3000 B.C.E., copper smelters in the Timna Valley, to the north of the Gulf of Eilat, Israel, were adding iron ore to crushed malachite and chalcocite copper ores. They had found that adding iron ore made the copper ore melt more easily and the melted material more fluid. A substance added to an ore to facilitate smelting is called a *flux*. The flux reacts with impurities in the ore, and the products of the reaction float on top of the molten metal as a *slag*. The Timna smelters introduced the technique of "tapping," by tilting the container and pouring off the slag while allowing the copper to accumulate at the bottom. When the slag cooled, metallic iron would have been found mixed with the other ingredients. In the 1930s the American archaeologist Nelson Glueck (1900–71) suggested that Timna Valley might have been the site of King Solomon's Mines, but this has not been subsequently verified.

Once the iron has been separated from its ore and cooled, it will contain a certain amount of carbon, which is absorbed from the charcoal used in the furnace. If the carbon content is low, a blacksmith can work the iron into the most elaborate forms. The result is wrought iron. The Iron Pillar in Delhi, India, erected in about 400 C.E. by the ruler Kumaragupta I in honor of his father, is more than 23 feet (7 m) tall and weighs more than 6.7 tons (6 t). It was not until the 19th century that European metalworkers were capable of handling a piece of metal that size. If the carbon content is higher, the iron can be melted and poured into a mold, producing cast iron.

Iron proved to be plentiful and extremely useful. It replaced bronze as the metal used to make weapons and tools in about 1200 B.C.E. in the Middle East and southeastern Europe and later in other parts of the world. That change marked the commencement of the Iron Age.

As the popularity of iron increased so did the demand for fuel to smelt it and to fire the blacksmiths' forges. That fuel was charcoal, and to maintain a steady and adequate supply landowners who supplied it managed their woodland by *coppicing*. Coppicing is the practice of cutting a tree close to ground level in order to encourage the growth of thin poles from around the edge of the stump. The diagram illustrates what happens when a tree is cut close to ground level. Vig-

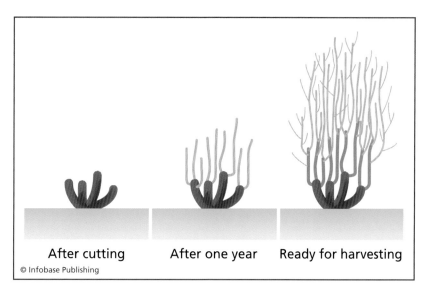

Coppicing, showing the silhouetted tree immediately after cutting, after one year, and when the poles are ready for harvesting

After cutting   After one year   Ready for harvesting

© Infobase Publishing

orous new growth has appeared by the end of the first year and after 12–15 years the poles are ready for harvesting. Trees were sometimes cut just below ground level, with a similar result. Most broad-leaved trees can be managed in this way; indeed, it greatly extends their lifespan. Coppicing was certainly being practiced by late in the Neolithic period (New Stone Age), before the widespread use of bronze, so the technique was far from new during the early Iron Age.

By the Middle Ages coal was taking the place of charcoal, and its smoke was making many European cities unpleasant places to be. But the real challenge to the metalworkers came with the change from iron to steel.

IRON INTO STEEL

Smelting iron calls for high temperatures, which can be achieved by forcing air through the burning fuel. Bellows were being used to force a draft by about 1800 B.C.E. The most basic furnaces are known as *bloomeries*, and the material they produced is called bloom.

The workers would choose the purest iron ore they could find, mix it with charcoal, and heat it strongly for several hours. The bellows would ensure a high temperature, and workers would add more fuel as required. The iron would form a spongy mass—the bloom—and carbon from the charcoal, ash from the fire, and other minerals from the ore would form a mixture called slag. Slag permeated the bloom. The furnace was then broken open and the white-hot bloom removed, and while still hot, it was hammered to expel as much of the slag as possible. Hammering also converted the bloom into a mass of wrought iron.

Iron has many uses, but it is too soft and brittle to make useful tools. By about 1400 B.C.E. East African metalworkers had discovered a way to overcome the difficulty. They managed the bloomery in such a way as to allow a little more carbon to remain in the bloom. They had produced the first steel. Steel is iron alloyed with up to about 1.7 percent carbon and certain other elements in modern steels; for example, stainless steel is steel containing 11–12 percent chromium.

Celtic peoples were expert metalworkers, and when Roman armies invaded the land they called Hispania (modern Spain and Portugal),

A ceremonial steel sword is held by a Guards officer during the Trooping the Colour ceremony, at Horse Guards Parade, London. *(Getty Images)*

the soldiers were impressed by the sharp steel swords used against them. The swords were curved, concave near the hilt and convex near the tip, with a single cutting edge. The smiths prepared the steel for them by burying steel plates in the ground and leaving them there for up to three years. During this time the weaker steel would rust away. They made the swords from the steel that survived. Swords are now obsolete as weapons of war, but soldiers still carry them for ceremonial purposes and fencing continues as a sport.

The Chinese were making steel a little more than 2,000 years ago but by a different method. They melted wrought iron and cast iron together. Wrought iron contains about 0.4 percent carbon, and cast iron 2–6 percent, so mixing them was a way to increase the carbon content.

By about 300 B.C.E. Indian and Sri Lankan metalworkers were making *wootz steel.* Their technique became widely used in the Middle East where, from about 1100 to 1700 C.E., it was called Damascus steel and renowned for making blades that were very tough and very sharp, with a distinctive appearance.

The wootz tradition died out in about 1700 because the ores it required had become difficult to obtain. Historians of metallurgy believe the steel was made by combining iron ore or wrought iron with charcoal and glass in a porous crucible. The crucible was sealed and heated slowly in a furnace, then cooled. This altered the contents of the crucible to a slag of impurities mixed with glass and buttons of metal containing about 1.5 percent carbon. These were forged into ingots. The finished result was a matrix of *martensite* or pearlite mixed with microscopic particles of iron carbide (Fe_3C). Martensite is a solid solution of carbon in iron with a particular crystal structure, known as alpha-iron. Pearlite is a mixture of iron with two different crystal structures. The carbide formed sheets and bands within the metal, giving the steel its characteristic appearance.

Although the process was producing steel in India and Sri Lanka around 2,300 years ago, crucible steel was first made in Europe in the 18th century. The process was invented by the English clock and instrument maker Benjamin Huntsman (1704–76). Huntsman needed a steel that was tough enough for clock springs. He was born at Epworth, Lincolnshire but opened his first business in Doncaster, Yorkshire. In 1740 he moved his operation to Sheffield to be closer to a source of fuel that was cheaper than charcoal.

That fuel was coke. Charcoal was made from coppice wood, and landowners maintained large areas of coppiced woodland to supply the iron industry, but as the demand for iron and steel increased, the supply became inadequate and the price rose.

It was not the needs of the iron and steel industry that led to the invention of coke, however, but the fact that ordinary coal (bituminous coal) could not be used to roast malted barley grains for brewing because the sulfur in the coal smoke would taint the resulting beer. Some other alternative to charcoal was called for. Charcoal is made by heating wood at about 750°F (400°C) under airless conditions. Coke is made by heating bituminous coal in the same way, at a temperature of about 1,800°F (1,000°C). The first coke was made in 1642 and used in Derbyshire, England, for roasting malt.

In 1709 the English ironmaster Abraham Darby (1677 or 1678–1717) introduced coke as a fuel in steelmaking. Coal was widely used in domestic fires and some manufacturing processes, but it was unsuitable for iron smelting and steelmaking because, despite several attempts to make it work, sulfur and phosphorus present in bituminous coal contaminated the iron. The process that converted coal to coke drove off moisture and all the volatile ingredients of the coal, leaving behind an almost pure form of carbon. Coke suitable for foundry work is about 90 percent carbon, and its high carbon content means it burns at a higher temperature than coal. It also possesses another important property. Coal expands as it is heated in the coking process, giving coke a strong, porous structure. The weight of ore does not crush the coke, and its porous structure allows air to circulate freely, ensuring an even heating of the charge.

HENRY BESSEMER—AND HIS CONVERTER

Until the early 18th century steel was made by the *cementation* process. This consisted of packing bars of wrought iron and charcoal in a clay box and heating it for several days, during which time the iron absorbed carbon and became *blister steel*. The process worked, but it was costly, because it took up to three tons of coke to produce one ton of steel. The crucible process, invented in Britain by Huntsman (see "Iron into Steel," pages 79–81), refined the cementation process, resulting in steel of a higher quality, but it added to the cost. Blister steel was broken into pieces and melted in the crucible, adding three hours—and much more coke—to the production process. In about 1865 about 280,000 tons (254,000 t) of blister steel and crucible steel were being produced in Europe annually, but during that same decade everything changed.

Cementation and crucible steel production were based on making iron absorb the desired amount of carbon. Cast iron had to be converted to wrought iron, after which carbon had to be added to the wrought iron to produce steel, which has a carbon content intermediate between the two. As the demand for iron had risen, larger furnaces had been introduced. These required a much stronger blast of air to drive the combustion process upward through the charge. The pressurized air produced a spongy mass of iron high in the furnace, and as the iron sank downward, it absorbed carbon from the fuel. The iron flowed from the base of the furnace along channels leading into molds for ingots. The ingots looked a little like piglets suckling from the sow, and the iron was called *pig iron.* The pig iron was then processed to make wrought iron.

The change that revolutionized steelmaking in the 1860s reversed the traditional process. Instead of adding carbon to iron, it removed surplus carbon, and it did so by oxidizing it, converting pig iron directly to steel in a device appropriately called a converter. The conversion process took about 30 minutes and produced steel for about £7 per ton (the equivalent of about £470 in today's money; about $740 per U.S. ton at the 2008 exchange rate), compared with up to £60 for a ton (about £4,000 in today's money; equivalent to $7,100 per U.S. ton) of crucible steel. (1 British ton = 2,240 pounds; 1 U.S. ton = 2,000 pounds.) The diagram shows how the converter worked. A charge of molten pig iron, usually about 17 tons (15 t), was placed in a steel container, approximately 10 feet (3 m) in diameter and 20 feet (6 m) tall, lined with either clay or dolomite, depending on the composition of the iron. The container was held by trunnions on either side, allowing it to be tilted to accept the charge and for the molten steel to be poured out. Air was blown under high pressure into the base of the container through channels called tuyeres. *Spiegel,* also called *spiegeleisen,* was added to the molten iron. This is a type of pig iron containing 15–30 percent of manganese and 4–5 percent of carbon. The manganese and carbon oxidized, removing surplus oxygen that would otherwise have given the steel a frothy texture.

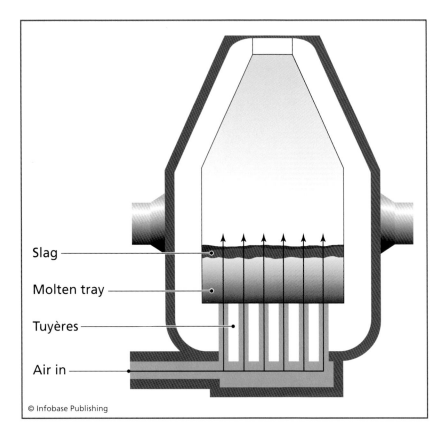

Slag

Molten tray

Tuyères

Air in

© Infobase Publishing

A Bessemer converter in which a blast of air, entering through tuyeres, blows through the molten iron in the molten tray, oxidizing impurities that accumulate as a slag

The converter is also known as the Bessemer converter, after the English inventor credited with designing and making it. In fact, though, Bessemer was not alone. William Kelly (1811–88), an American ironmaster who was born in Pittsburgh and died in Louisville, Kentucky, discovered a very similar process at about the same time. Kelly had been experimenting since about 1850 with ways to make steel directly from pig iron. He finally succeeded and patented his process, and the Kelly Pneumatic Process Company commenced production in 1863 but quickly found itself competing with a company using the Bessemer process. The two companies merged, steel production increased rapidly, and in the United States the method they used became known as the Kelly-Bessemer process.

Henry Bessemer (1813–98) belonged to a French Huguenot family who had moved to England at the time of the French Revolution

(1787–99), and Bessemer was born on January 19, 1813, at the village of Charlton, near Hitchin in Hertfordshire, to the north of London. His father worked for a company making metal printing type, and Henry spent a great deal of his time in his father's workshop, where he developed his talent as an inventor. He invented a new method for stamping deeds, a way of making blocks for printing artwork that led to him being invited to exhibit at the Royal Academy, a typesetting machine, and a process for making imitation velvet. He also invented a method for making bronze powder and gold paint, used in printing, and then turned his attention to improving methods for refining sugar. He even invented a new way to make "lead" pencils. His bronze powder and gold paint were at least as good as those being imported at the time from Germany, and they were much cheaper. The profits from this enterprise allowed him to open a small ironworks in St. Pancras, London.

In the early 1850s Britain and France were at war with Russia in the Crimea, and Bessemer became interested in guns. Gunmakers were experimenting with rifling. Rifling (which gives the *rifle* its name) is a spiral flange inside the barrel that sets the shell spinning, making its flight more stable and therefore increasing the range and accuracy of the gun. The difficulty was that in order to work, the shell had to fit tightly inside the barrel, and this meant the barrel was subjected to very high pressure from the explosion of the charge. Guns tended to explode, killing the gunners. Bessemer offered his services to the British military, but they were not interested so he approached the French, who were, and Napoléon III encouraged Bessemer to continue with his experiments aimed at producing a steel strong enough to make rifled gun barrels.

Bessemer discovered that forcing a blast of air through molten iron oxidized impurities but also raised the temperature of the metal, greatly reducing the amount of coke that was needed. He announced his discovery in 1856 at a meeting of the British Association for the Advancement of Science. Ironmakers were enthusiastic and invested heavily in the new "blast furnaces," but their enthusiasm evaporated when they found the steel they were making was of very poor quality, and Bessemer was laughed at. It transpired that phosphorus prevented the process from working. Bessemer had used phosphorus-free ore to make pig iron, but the ironmakers who adopted the process used ore containing phosphorus. Bessemer explained this, but

A photograph of a Bessemer converter at work *(Hulton Archive/Getty Images)*

it was too late; he had lost all credibility. So he established his own steelworks in Sheffield, Yorkshire, making steel from phosphorus-free ore he imported from Sweden. Soon he was producing steel more cheaply than his competitors, and they rushed to buy licences to install his patented process. Steel prices fell, and Henry Bessemer became extremely wealthy. While the drawing on page 83 shows how his converter worked, the photograph of one in use shows what it really looked like—and gives an impression of the intense heat and noise in which men had to work.

Henry Bessemer was elected a fellow of the Royal Society in 1879, and the same year he was knighted for his service to the Inland Revenue (equivalent to the Internal Revenue Service) in inventing the method for stamping deeds. The recognition was well deserved but late in coming: He had made his stamping invention 40 years earlier, before he was 20. The French government wanted to award him the Legion of Honor, but the British authorities forbade him to accept it. Sir Henry Bessemer died in London on March 15, 1898.

The Bessemer converter was a huge success, not least because it could make steel suitable for rail tracks and wheels, which until then had been made from iron. In 1867, for example, 460,000 tons (418,000 t) of iron rails were made in the United States and sold for $83 per ton ($75/t), and 2,550 tons (2,315 t) of Bessemer steel rails, which sold for up to $170 per ton ($154/t). The last iron rails were made in 1884. Steel rails had taken their place, and in that year U.S. steelworkers produced 1.5 million tons (1.4 million t) of them. They sold for $32 per ton ($29/t).

Improvements continued to be made. In 1876 the English chemists and metallurgists Sidney Gilchrist Thomas (1850–85) and his cousin Percy Carlyle Gilchrist (1851–1935) found a solution to the phosphorus problem. If the converter was lined with limestone or dolomite and burned lime was added to the pig iron charge, the slag would trap and hold the phosphorus. Steel made this way was called Thomas steel.

Eventually the Bessemer process fell from use. Newer ways of producing steel were able to accept scrap steel and high-phosphorus ores, and they allowed greater control over the chemistry of the production process. These had largely replaced the Bessemer process by 1900, and no Bessemer steel has been made in the United States since 1968. Like most machines, when they became obsolete, Bessemer converters were broken up for scrap. There are now very few still in working order anywhere in the world.

AGRICOLA—AND THE FORMATION OF ORES

Most metals are smelted from ores, which are rocks containing a chemical compound of the desired metal in a concentration high enough to allow it to be extracted and purified commercially. Rocks are made from minerals. Metal ores are minerals, but not all minerals are metal ores. Some can be cut and polished to make ornaments, and if they are rare enough, attractive minerals become precious stones from which jewels are fashioned. Gemstones often owe their appearance to impurities in an otherwise commonplace mineral. Corundum, for example, is an aluminum oxide (Al_2O_3) that is very hard (hardness 9 on the Mohs's scale, which measures the hardness of minerals from talc at 1 to diamond at 10). It is used to make grinding wheels and emery paper, but impurities can color and alter it. The impure forms of corundum include sapphire, ruby, and emerald.

Miners must learn to recognize rocks that contain minerals of interest or value, and in practical terms nowadays this means minerals that can be processed into metals. It was not always so. In the Middle Ages many minerals were believed to possess therapeutic properties, and physicians made use of them. Miners also needed to recognize medicinal minerals.

Scholars, too, were interested in rocks and minerals, and in the middle of the 16th century Georgius Agricola published two books that earned him the title "father of mineralogy." Agricola's observations, ideas, and the drawings of mining machinery with which he illustrated his books were of such high quality that they became standard textbooks and reference works. Both books remained in use for centuries, and they are still respected by scientists. Agricola wrote in Latin, and in 1912, before he became the 31st president of the United States, Herbert Hoover and Mrs. Hoover (Lou Henry) translated into English his book on mining and metal production, *De re metallica* (About metals). Hoover was an authority on mining and metallurgy and Lou Henry was a geologist.

Agricola was not the author's real name. It was usual for scholars to use Latin names on the title pages of their literary works, a fashion that continued until the 18th century. Georgius Agricola was born

Georg Bauer. *Bauer* is the German word for "peasant" or "farm-worker," so *Agricola* (farmer) was a straightforward translation.

Bauer was born on March 24, 1494, at Glauchau, in the German state of Saxony. He enrolled as a student at the University of Leipzig in 1514, graduating in 1518 with a bachelor's degree in classics and philosophy. After leaving university he taught Latin and Greek at the Great School in Zwickau, a town to the south of Leipzig, remaining there until 1522 and becoming principal. Then he returned to Leipzig, but in 1523 he moved to Italy to continue his studies at the universities of Bologna, Venice, and possibly Padua. In Italy he studied medicine, natural history, and philosophy, qualifying as a physician in 1526. He returned to Saxony the same year and took up an appointment as town physician in St. Joachimsthal, a town that had been founded only a few years earlier and had a population of about 1,000 persons. Revision of the national frontier means the town is now in Bohemia, in the Czech Republic, and it is called Jáchymov.

The move was important in Bauer's life. Bohemia was an important mining region, and silver had recently been discovered near St. Joachimsthal. While he was in Italy, Bauer read *Utopia*, a story (sometimes described as the first novel) by Sir Thomas More (1478–1535), published in Latin in 1516, describing a land ruled entirely by reason. The book inspired Bauer to study science, and the aspect to which he devoted himself centered on the mining industry around him. His initial interest probably arose from the medical uses of minerals. He published a short book on mining and mineralogy in 1530 and then resigned his position to spend time traveling and studying mining and the lives of miners.

In 1530 Prince Maurice of Saxony (1521–53) appointed him historiographer (a scholar who studies history and historical methods) with an annual allowance, and he moved to Chemnitz, Saxony, where in 1533 he was elected town physician. He was familiar with the health problems experienced by miners and treated them on the basis of his own observation and understanding rather than following traditional medical practice. Bauer was the first doctor to introduce quarantine in Germany.

Mines and mining were his real love, however, and he devoted all his spare time to them. He wrote a number of books, but only two have ever been translated from Latin. The first of his major works was *De natura fossilium* (The nature of fossils), published in 1546. Bauer was the first person to use the word *fossil* to describe anything that was dug from the ground. The Latin *foss-* means "dig," so this is still the literal meaning of fossil; it is why peat, coal, and oil are called "fossil fuels." *De natura fossilium* was the first comprehensive classification of minerals based on Bauer's own observation of their geometric forms and to some extent their composition, although little was known then about the chemical composition of minerals. Bauer included a long and thorough review of the existing literature on the subject but always relied on his own observation and interpretation. This was the book that established Bauer's reputation.

His second book was published posthumously in 1556. *De re metallica* described mining and mineralogy. It reviewed the existing literature and described mine ownership and mining laws in Saxony, but the main part of the book dealt with the geology of ore rocks, the assaying of ores, smelting and ore enrichment, and the construction and management of mines. It was lavishly illustrated with woodcuts and included a glossary. This was necessary, because Bauer had to invent Latin words for the many German scientific and technical terms he used.

In 1546 Bauer became mayor of Chemnitz. A Catholic, he participated in negotiations between the Protestant states of Germany and the Holy Roman Emperor. Bauer married, possibly twice, and had several children. For most of his life, Bauer's staunch Catholicism was tolerated, but in the 1550s Protestantism grew increasingly militant in Germany, and Bauer's popularity waned. He died on November 21, 1555, at Chemnitz, but anti-Catholic feeling was running so high that he could not be buried in the town. Instead, amid hostile demonstrations, his body was taken to Zeitz, a town about 30 miles (50 km) away, and buried there.

ALBERT THE GREAT—AND THE SCIENCE OF MINERALS

Modern European science began when scholars first gained access to the works of Greek philosophers, especially Aristotle (384–322 B.C.E.). The original Greek works were translated first into Arabic by Islamic scholars who used them as a basis for their own work, especially in mathematics, astronomy, and medicine. The Arabic writings were then translated into European languages, most of them entering Europe from Spain, then under Moorish occupation. Philosophers and what today would be called scientists then had to read the ancient works critically. One of the most eminent of these early critics was St. Albertus Magnus (ca. 1200–80).

Albert was the eldest son of the count von Böllstadt, a title he later inherited. He was born at Lauingen an der Donau, Swabia (Swabia covered territory that now extends across the borders of Germany, Switzerland, and Alsace, France) some time between 1193 and 1206. He was probably educated in Lauingen before enrolling at the University of Padua, where he first encountered Aristotle's writings. After graduating, in 1223 he joined the Dominican Order and studied and taught at Padua, Bologna, Cologne, Hildesheim, Freiburg im Breisgau, Ratisbon (now Regensburg), Strasbourg, and other places, staying in Dominican houses. In about 1241 he went to Paris, where he taught for four years. He received a master's degree in theology from the University of Paris in 1245. One of his first students after that was Thomas Aquinas (ca. 1225–74).

It was while he was in Paris that Albert began writing his commentaries on every aspect of knowledge acquired from the Greeks, adding his own observations. In his *De mineralibus* (About minerals) he wrote: "The aim of natural science is not simply to accept the statements of others, but to investigate the causes that are at work in nature." This was a radical statement, for in those days most people believed that a careful study of scripture was the only way to knowledge. Albert had proposed a skeptical, inquiring approach that today is the foundation of the scientific approach to natural phenomena. Nevertheless, he was able to present this in a way the church could accept. Albert was critical of Aristotle and of those who accepted Aristotelian ideas without questioning them, but that did not prevent his enemies from calling him "Aristotle's ape." His fellow scholars called him Albertus Magnus (Albert the Great) in recognition of his learning. He was also called Doctor Universalis (Universal Doctor).

Albert left Paris in 1248 to set up a Dominican university in Cologne, and he remained there as rector until 1254, when he was appointed superior (head) of the Dominican province of Teutonia (Germany). This position demanded more of his time and attention than he wished to divert from his scientific work, and in 1257 he resigned and returned to Cologne. In 1260 he was appointed bishop of Ratisborn but again resigned in 1262 to return to Cologne. He died in Cologne on November 15, 1280.

Albert was beatified in 1622, and on December 16, 1931, Pope Pius XI declared him to be a Holy Doctor of the Church. This was automatic canonization, and Albert is one of only 33 persons to be honored in this way. In 1941 Pius XII made him patron saint of natural scientists. His feast day is on November 15.

What Are Fossils?

Certain rocks contain sections that closely resemble seashells but shells that are made entirely from stone. Rocks may also contain what appear to be the crushed and distorted skeletons of strange-looking animals or black impressions of fish unlike any known today. Coal contains the shapes of insects and plants, including the apparent impressions of leaves, fern fronds, and tree bark.

People have always known of these curious rocks. They collected them and made jewelry from those they were able to separate from the rock holding them. Perhaps fossils were used in religious ceremonies. Fossils have been found in some Neanderthal graves, objects of importance placed there by members of *Homo neanderthalensis,* a now-extinct species related to but different from modern humans.

Scientists can make educated guesses about the uses prehistoric peoples found for fossils, but they can only speculate about how those people explained their existence, if they even attempted such a task. Rocks contained other attractive, interesting, or valuable objects: Gold nuggets could be prized from some rocks, copper from others, and gemstones also lay concealed inside rocks. By the Middle Ages anything dug from the ground was described as a fossil, so these strangely formed stones were but one type of fossil among many. When it became fashionable for wealthy individuals to store intriguing objects in their "cabinets of curiosities" and invite their friends to view and admire them, the cabinets usually contained fossils of these different kinds. Some of the biggest of the private collections of curiosities eventually formed the basis of museum collections.

Many collectors believed that the strangely shaped stones had simply formed that way within their rocks, rather in the way that cultivated plants occasionally gave rise to "sports," which are offspring of a different form or color. They were curiosities, but nothing more. Others thought Satan had made and placed them inside rocks. In China fossil bones and teeth were thought to have belonged to dragons, and ground into powder, they were used in traditional medicine.

Theophrastus (ca. 372–ca. 287 B.C.E.) was possibly the first person to recognize that fossils resembling animals and plants were actually linked to living organisms (see sidebar). So far as anyone knows, however, he did not suggest that they were the actual remains or traces, greatly transformed, of organisms that had once lived. He thought they had begun to form by *spontaneous generation.* This is the theory, finally disproved in the late 17th century but persisting into the 19th century, that living organisms can emerge directly from nonliving matter. This is what Aristotle believed, and Theophrastus was his pupil.

The Arab scholar Ibn Sīnā, known in Europe as Avicenna (979–1037), proposed a modified version of this idea. He suggested there are "plastic forces" or "formative virtues" operating inside the Earth's rocks. Fossils were either organisms that had started to form by spontaneous generation but had failed, or they were the result of "vital essences," possessed by all living things, that had penetrated rocks and formed objects shaped like the organisms from which these "vital essences" had come.

It was a long time before there was a serious challenge to these ideas. When scientists finally accepted that fossils provide a record of once-living organisms and the discipline of *paleontology* was born, the scientific understanding of the Earth changed radically. The Earth had acquired a history extending an unimaginably long distance into the past.

KONRAD VON GESNER—AND HIS FOSSILS

Christophorus Encelius (1517–83), a German naturalist, included four drawings of fossils in a book he published in about 1551. These are believed to be the first scientific illustrations of fossils, but they

THEOPHRASTUS, WHO CLASSIFIED MINERALS AND WROTE ABOUT FOSSILS

A Greek philosopher who wrote more than 200 books on a wide range of topics, Theophrastus (ca. 372–ca. 287 B.C.E) is best known as the founder of the science of botany. He classified plants as trees, shrubs, undershrubs, and herbs; made detailed studies of flowers; and distinguished between flowering and cone-bearing plants and between monocots and dicots, the two main groups of flowering plants.

Theophrastus also classified minerals and wrote about fossils. His book on fossils is lost, but he referred to it in another of his books, *Peri lithon* (*About Stones*). It is clear from this that he understood that fossils are related to living organisms rather than being stones bearing a curious but coincidental resemblance to them. In this he was clearly correct, but he accepted the Aristotelian view that living organisms can emerge from non-living substances, a theory known as spontaneous generation, or abiogenesis. According to this theory, which prevailed among scientists until the late 17th century, living organisms would appear wherever the necessary prerequisites occur. According to this view, which is probably what Theophrastus believed, fossils were organisms that had arisen spontaneously but had failed to emerge from the rocks in which they had formed.

Theophrastus was born at Ephesus, on the Greek island of Lesbos. His original name was Tyrtamus. He studied at the Academy in Athens under Plato (ca. 427–ca. 347 B.C.E.), where Aristotle (384–322 B.C.E.) was also working. It was Aristotle who gave Tyrtamus the nickname *Theophrastus*, which means "divine speech." When Aristotle left the Academy following Plato's death, Theophrastus accompanied him and became his chief assistant when Aristotle returned to Athens in 335 and established his own school, the Lyceum. When Aristotle left Athens in 322, Theophrastus became head of the Lyceum, which prospered under his direction. Theophrastus remained there until his death in about 287.

were greatly surpassed in a book by Konrad von Gesner (1516–65) under the full Latin title *De Rerum Fossilium, Lapidum et Gemmarum maxime, figuris et similitudinis Liber: non solum Medicis, sed omnibus rerum Naturae ac Philogiae studiosis, utilis et juncundus futurus.* In English it is known as *Fossils, Gems, and Stones.* Gesner's book, published in 1565, contained many illustrations—including those from Encelius's book—and a classification for them.

Konrad Gesner was a Swiss physician, botanist, zoologist, and naturalist. He was born in Zurich, Switzerland, on March 26, 1516, the son of a furrier. Gesner began his education in Zurich. The family was not wealthy and became very poor after Konrad's father was killed in a battle in 1531, but friends and patrons helped the young man, making it possible for him to study at the Universities of Strasbourg and Bourges. Religious conflicts forced him to return to Zurich in 1535. He continued his studies at the University of Basel in 1536, and in 1537 he became professor of Greek at an academy recently founded in Lausanne. In 1540 he went to the University of Montpellier, in France, to study medicine, received his degree as a doctor of medicine at Basel in 1541, then settled in Zurich, where he was elected town physician. In 1564 Gesner was ennobled, adding the *von* to his name. In the following year, 1565, there was an outbreak of plague in Zurich. Gesner refused to abandon his patients, and when he fell ill, he asked to be carried to his study. There, on December 13, Konrad von Gesner died, surrounded by his books and collection of curiosities.

A polymath is a person who knows about a wide range of subjects. Gesner was a polymath who devoted every moment he could spare from his patients to accumulating information, organizing it, and communicating it to others. He produced about 90 manuscripts, on zoology, theology, medicine, mountains, plants, and many other topics. His zoological work, *Historia animalium* (Story of animals), consisting of four volumes published in Zurich between 1551 and 1558 and a fifth published in 1587, is considered one of the starting points of modern zoology. He also wrote an account of approximately 130 languages, with the Lord's Prayer in 22 languages. His most remarkable work, *Bibliotheca universalis* (Universal library), appeared in 1545. This was nothing less than a catalog of approximately 1,800 writers with a list of the titles of all their books and a summary of each, written in Latin, Greek, and Hebrew.

Gesner devised a method for classifying fossils, dividing them into 15 groups, but along lines that now seem eccentric. His class 3, for example, grouped together echinoderms (a phylum of invertebrate animals including sea urchins and starfish), Neolithic stone axes, and minerals with a smoky appearance, all of which he believed were objects that had fallen from the sky (an idea originally advanced by Aristotle). Gesner's class 13 comprised gems, minerals, rocks, and fossils with names derived from birds.

Gesner pointed out the similarity between certain fossils and living organisms: His class 10 contained fossils resembling corals, 11 looked like sea plants, 14 like marine animals, and 15 like insects or snakes. Gesner did not however suggest that these objects were

Fossil of the trilobite *Trevoropyge prorotundifrons*, which lived from the Lower Ordovician to Upper Devonian (488–359 million years ago). This fossil was found in Morocco. *(Sinclair Stammers/Science Photo Library)*

directly derived from living organisms. Classifying fossils in this way cannot have been simple, because there are some that bear very little resemblance to any organism alive today. Trilobites, for example, were arthropods, the phylum that includes crustaceans, insects, and arachnids, but the three-lobed body plan that gives them their name makes them unlike any modern animal. Some could swim, but most lived on the seabed. The photograph shows a fossil trilobite (*Trevoropyge prorotundifrons*) found in Morocco. Not all fossils are so well preserved, but even this fossil is quite unlike any existing animal.

It was not only fossils that were generally misunderstood at that time. Archaeology had not yet begun, and scholars had no knowledge or means of acquiring knowledge of the way people lived in prehistoric times. Large standing stones and stone circles such as Stonehenge were thought to be the work of a race of giants that had long since disappeared. People believed that stone objects discovered below ground had either fallen from the sky or had been formed like other rocks, their forms a curiosity but with no deeper significance. Stone arrowheads and points were sometimes called "elf arrows," and bigger objects were "thunderstones." Gesner called them *ceraunia,* a term that later came to include all hard objects dug from the ground.

Gesner was enthusiastic, energetic, and what would nowadays be called a workaholic, but he was not an innovator. Yet even during his lifetime there were some who realized that the close similarities between fossils and living organisms or parts of them could not be coincidence. If a fossil resembled a leaf, footprint, seashell, bone, or tooth, then perhaps that is exactly what it once had been.

LEONARDO DA VINCI, WHO SAW FOSSILS FOR WHAT THEY ARE

Even before Gesner was born, the Italian painter and inventor Leonardo da Vinci (1452–1519) had recognized in the late 15th century that fossil shellfish were genuinely the remains of animals that had once been alive and that their presence in rocks on dry land was evidence that in ancient times what is now dry land had lain beneath the sea. But Leonardo kept his knowledge to himself, recording it in notebooks written during the last 30 years of his life. In his notebook Leonardo used abbreviations, unusual spellings, and mirror-writing (writing backwards and from right to left) to make them difficult to read, and purposely did not arrange the information in any coherent order. Historians believe Leonardo intended to publish them as a book, but this was one of many of his projects that he failed to complete. The 4,000 surviving pages of his notebooks on scientific and technical subjects were not published until the 19th century.

Leonardo worked from 1482 until 1499 for Ludovico Sforza, the duke of Milan, as a "painter and engineer of the duke." His duties included planning major civil engineering projects that would drive tunnels through mountains and even remove mountains altogether. In the course of this work he made many observations of mountains and rivers, realizing that sediments transported by rivers can be compressed to form rocks and that rivers erode rocks. He observed that sedimentary rocks consist of layers, each deposited on top of an earlier layer, and he found that layers of rock containing characteristic fossils covered large areas but were formed at different times.

It was well known that rocks on the tops of mountains contained fossil seashells. Some people thought these stones might have grown inside the rocks, as proposed by Avicenna. Leonardo dismissed this idea, writing (in his notebooks) that "such an opinion cannot exist in a brain of much reason; because here are the years of their growth, numbered on their shells, and there are large and small ones to be seen which could not have grown without food, and could not have fed without motion—and here they could not move." Others suggested that the biblical flood had deposited the shells, but Leonardo would have none of that idea, either. He said the water would have had nowhere to go when the flood receded, and if the shells had been transported by torrents of water, they would have become thoroughly mixed and distributed throughout the mud, not deposited in the regular layers in which they were found. He also pointed out that water flows downhill, not uphill, and would have carried the shells away from the mountains, not toward their summits. He could not imagine how such a flood could have carried oysters and corals 300 miles (480 km) inland, or how these organisms could have crawled such a distance during the 40 days and nights the Flood was said to have lasted.

So, what were these stones? Leonardo thought they were once-living organisms that had been buried in sediments before those sediments were lifted to form mountains. Some rocks where fossils were found had formed along sea coasts, and some sediments were deposited by floods, but by local floods, not a single, worldwide flood. Leonardo realized that fossils record the history of the Earth extending back far beyond any written historical record. "Since things are much more ancient than letters," he wrote, "it is no marvel if, in our day, no records exist of these seas having covered so many countries."

Leonardo was born on April 15, 1452, in Vinci, Tuscany, in Italy, the illegitimate son of the landowner Ser Piero and a peasant girl. He was taken into his father's household in Florence. After receiving an elementary education, in about 1467 he was apprenticed to an artist, Andrea del Verrocchio, learning about technical and mechanical matters as well as painting. He left Andrea's workshop in 1477 and in 1482 went to work for the duke of Milan. In 1499 the French captured Milan, and Leonardo fled. He visited Mantua and Venice in 1500, and in 1502 he began work as a military architect for Cesare Borgia. He then moved to Milan in 1506 to work for the French governor, Charles d'Amboise. When the French were forced to leave Milan in 1513, Leonardo sought work in Rome. In 1516 the French king, François I, invited him to France. He accepted and spent the rest of his life at the castle of Cloux, near Amboise, where he died on May 2, 1519. The castle is now a museum containing wooden models of machines based on his drawings, and visitors can see the kitchen where, as a very old man, he would sit on cold winter days.

Although he is best known as a painter, Leonardo possessed a wide range of talents. As an inventor and mechanical engineer he drew designs for flying machines, an armored tank shaped like a tortoise, pontoons, a device for breathing underwater, a boat propelled by paddlewheels, drilling machines, lathes, and a clock with a minute hand as well as an hour hand. He designed and is believed to have built the first elevator, in Milan's cathedral. He studied human anatomy and speculated about the circulation of the blood. He also studied rocks and fossils. Perhaps inevitably, his wide range of interests and his perfectionism meant that he failed to complete many of the projects he started. His notebooks on painting were published in the 17th century, but it was not until the translation and publication of his other notebooks two centuries later that the scale of his genius came to be recognized.

Leonardo's recognition of the true nature of fossils represented a radical break from earlier ideas. It was not simply that Leonardo noted something others had failed to observe; he positively rejected an entire mode of thought, an entire mindset. This may not have been deliberate on his part but the result of his practical, down-to-earth outlook. He was a painter, inventor, and engineer rather than a philosopher.

The mindset he broke derived from Neoplatonism, a philosophical outlook founded by the Greek philosopher Plotinus (204–270 C.E.). Plato (ca. 427–ca. 347 B.C.E.) had taught that all the objects and phenomena in the world are mere reflections of perfect versions of those objects and phenomena, which exist in a hidden ideal world. Plotinus developed Platonic ideas further, but in doing so, he changed them. Plotinus believed in a holy trinity consisting of One, Spirit, and Soul. These are not equal. One, which is a rather vague concept, is supreme, Spirit comes next, and Soul is last. Spirit (Greek *nous*) has a strong intellectual connotation. Soul springs from the Divine Intellect and permeates all living beings and nonliving objects. Neoplatonism strongly influenced the ideas of scholars throughout much of the 15th and 16th centuries. The Neoplatonists taught that "soul" permeates everything in nature, but that "soul" is inferior to and can be controlled by "spirit" in the sense of the intellect. If only the secret keys can be found, therefore, the human intellect will be able to understand everything. The Neoplatonists were more interested in and tolerant of magic and astrology than Plotinus himself had been, but by the time of Leonardo and Gesner these ideas and the symbolism derived from them were very important. The beautiful illustrations of plants, animals, and fossils produced at that time were meant to aid in the study of nature but with due regard to the symbolism of natural forms. In taking a more literal view, Leonardo was nearly two centuries ahead of the scholars of his own time.

ROBERT HOOKE, WHO SHOWED THAT LONG AGO BRITAIN LAY BENEATH THE SEA

Leonardo had dismissed Avicenna's idea of "plastic forces," but his notes had not been published, so Avicenna's view remained popular throughout most of the 17th century, until Robert Hooke (1635–1703) emphatically refuted them. In *A Discourse of Earthquakes*, published posthumously in 1705, Hooke wrote of fossils (he called them "petrifactions"):

> [I]f the finding of Coines, Medals, Urnes, and other Monuments of famous persons, or Towns, or Utensils, be admitted for unquestionable Proofs, that such Persons or things have, in former Times, had a being, certainly those Petrifactions may be allowed to be of equal Validity and Evidence, that there have been formerly such Vegetables or Animals. These are truly Authentick Antiquity, not to be counterfeited, the Stamps and Impressions, and Characters of Nature that are beyond the Reach and Powers of Humane Wit and Invention, and are true universal Characters legible to all rational Men.

In other words, humans lack the skill to have manufactured fossils and embedded them in rocks, so they should be accepted as the genuine remains of plants and animals.

Also in *A Discourse of Earthquakes* Hooke dealt with the problem of fossilized marine animals that are found on mountains, at high elevations and far from the sea. He wrote:

> . . . the Waters have been forc'd away from the Parts formerly cover'd, and many of those surfaces are now raised above the level of the Water's Surface many scores of Fathoms. It seems not improbable, that the tops of the highest and most considerable Mountains in the World have been under Water, and that they themselves most probably seem to have been the Effects of some very great Earthquake.

Hooke was the most talented experimenter in 17th-century England and arguably one of the greatest scientific geniuses Britain has ever produced. His interests were wide, he corresponded and collaborated with many other eminent scientists, and when he encountered a problem that could be solved by experiment but for which there was no suitable equipment, his response was always the same: He invented and built whatever was required.

Hooke was also a keen microscopist and excellent artist, who invented the compound microscope and its lighting system. His subjects included fossils—he was the first person to examine them under a microscope—and he compared fossilized wood with rotten wood and a fossil shell with a living shell. In the case of the rotten wood he actually observed the process of petrification and described it in *Micrographia*, published in 1665:

> [T]his petrify'd Wood having lain in some place where it was well soak'd with petrifying water (that is, such water as is well impregnated with stony and earthy particles) did by degrees separate abundance of stony particles from the permeating water, which stony particles, being by means of the fluid vehicle convey'd, not onely into the Microscopical pores, but also into the pores or Interstitia of that part of the Wood, which through the Microscope, appears most solid.

Hooke believed that seashells were fossilized by a similar process. As for "plastic virtue," he wrote:

> That these figured Bodies dispersed over the World, are either Beings themselves petrif'd, or the Impressions made by those Beings. To confirm which, I have diligently examin'd many hundreds of these figured, and have not found the least probability of a plastick Faculty.

Robert Hooke was born on July 18, 1635, in Freshwater, on the Isle of Wight (off the south coast of England). His father wished him to enter the church, but Robert was a sickly child and never strong enough to attend school. Instead his father educated him, but the boy was often alone and passed his time by making all kinds of mechanical toys. After his father's death in 1648, Hooke went to London. At first he paid the £100 he had inherited from his father to begin an apprenticeship, but soon he enrolled at Westminster School, where he studied Latin and Greek and was first introduced to mathematics. In 1653 he went to Christ Church College at Oxford University as a chorister, being allowed to attend lectures in return for singing in a choir, and later became a servitor, which is an undergraduate student who receives assistance from college funds in return for performing certain menial tasks. Already a talented instrument maker, Hooke supported himself by selling ideas for modifications and improvements to the owners of professional instrument workshops.

Hooke did not take a degree, but his skills attracted the attention of some of the most talented scientists of the day, and he became an assistant, and lifelong friend, of Robert Boyle (1627–91). Hooke left Boyle's employ in 1662 to become curator of experiments at the recently formed Royal Society, but Boyle continued to help him financially until 1664, the year Hooke was appointed lecturer in mechanics at the Royal Society with an annual salary of £50. In 1665 Hooke became Gresham Professor of Geometry at Gresham College, London. His salary there was also £50 a year, bringing his total salary to £100, as he continued to lecture at the Royal Society, and he was provided with rooms at the college, where he lived for the rest of his life. He was secretary to the Royal Society from 1677 until 1683. His health deteriorated over the last 10 years of his life, and he died in London on March 3, 1703.

By the end of the 17th century the tide had turned. The Neoplatonic explanation of fossils as stones that corresponded to living organisms had been rejected, and scientists recognized their true origin. Hooke contributed greatly to this change, his thorough research and forthright explanations undermining the old ideas. But perhaps the final blow was delivered by one of the most remarkable scientists of all time, and a man who is destined to become a Catholic saint.

NICOLAUS STENO, WHO FULLY UNDERSTOOD FOSSILS

Nicolaus Steno is the English version of Nicolaus Stenonis, the Latinized name of Niels Stensen, also spelled Steensen (1638–86), a Danish anatomist and geologist. Stensen was born a Lutheran but converted to Catholicism while in Italy. On the day he was accepted into the Catholic Church, Steno received a letter recalling him to Copenhagen to become royal anatomist, a position at the royal court. He was permitted to return to Protestant Denmark despite being a Catholic, but after two years he petitioned the king for permission to return to Italy. This was granted and Steno moved to Tuscany, where he became tutor to a prince of the Medici family. In 1675 Steno became a priest and took a vow of poverty. He was summoned to Rome in 1677, made a bishop, and appointed vicar apostolic with the task of converting Protestants to Catholicism in Northern Germany and Scandinavia. He took his vow of poverty very seriously, and his friends grew increasingly concerned for him as he grew pale and emaciated. Steno was 48 years old when he died at Schwerin, Germany, on November 25, 1686. He is buried in the Medici crypt in Florence. In 1988 Pope John Paul II beatified Steno, a preparatory step toward canonization (sainthood).

That is one side of Steno's life. He was also a leading scientist. Born in Copenhagen on January 1, 1638, he was the son of a prosperous goldsmith. He studied anatomy at the University of Copenhagen and medicine in the Netherlands, at the Universities of Amsterdam and Leiden. While in the Netherlands he discovered what is now called Steno's duct of the parotid gland and studied the way the ovaries work. In 1664 he moved to Paris, where he delivered a discourse on the anatomy of the brain that was published in 1669. In this he also showed that animals other than humans possess a pineal gland, thus disproving Descartes's idea that the pineal gland is the seat of the soul. He visited Montpellier, in France, and in 1665 he moved to Florence, where, in 1666, he was appointed personal physician to Ferdinand II (1610–70), the grand duke of Tuscany and an enthusiastic patron of science. But although Steno made major contributions to anatomical knowledge, his most important work concerned rocks, fossils, and the conclusions he drew from his studies of them. Steno studied crystals and discovered in 1669 that regardless of their shape, the angle formed by corresponding faces never varies for a particular mineral. This is now known as Steno's law.

His work on fossils all began when, in October 1666, two fishermen caught a large shark near the town of Livorno and sent its head to Duke Ferdinand. Ferdinand gave it to Steno, who dissected it. In doing so he was struck by the similarity between the shark's teeth and a class of fossils called *glossopetrae,* or tongue stones. These had been found in several parts of Europe, especially on the island of Malta, and had been known since ancient times. The stones were believed to possess magical healing powers. Their name referred to their shape, which resembled that of a human tongue. Pliny the Elder (23–79 C.E.), the Roman naturalist and writer, suggested that tongue stones fell from the sky, possibly from the Moon. Others believed the stones grew inside rocks. Steno, on the other hand, maintained that tongue stones had once been teeth inside the mouths of sharks. They had been buried in mud or sand on the seabed, and the seabed had subsequently been raised above sea level. He found that the stones were not made from the same substance as teeth, but he argued that their chemical composition could alter without changing their shape.

Steno was not the first to recognize tongue stones as shark teeth. Robert Hooke and John Ray had made a similar assertion, and in 1616 the Italian naturalist Fabio Colonna (1567–1650) explicitly stated as much in his *De glossopetris disertatio* (Dissertation on glossopetrae). However, Steno went further. He thought deeply about the problem of "solids within solids." How did these organic remains come to be sealed inside rocks, and how did layers of crystals or mineral veins come to be lying between layers of entirely different types of rock? He published his ideas on the subject in 1669, in his most famous work, often known simply as *Prodromus*, but with the full title *De solido intra solidum naturaliter contento dissertationis prodromus* (Preliminary discourse to a dissertation on a solid body naturally contained within a solid).

Steno proposed that many rocks and minerals had once existed as particles suspended in a liquid such as water, and they had precipitated from the liquid to form horizontal layers. The layers, or rock strata, form in a horizontal or nearly horizontal position, and if they are found oriented in any other way, it means they have subsequently been disturbed. This is now called the *law* (or principle) *of original horizontality.* He also stated that if a solid body were to be enclosed within another solid body, the surface of the second body to set hard would conform to the configuration of the surface of the body that hardened earlier. If tongue stones or other fossils had grown inside the rocks in which they were found, they would have been distorted by the solid rock around them. Instead, they must have been hard objects that were buried in soft sediment that was subsequently changed to hard rock.

As particles fall from suspension, they form layers of sediment one on top of another. Consequently, Steno argued, the oldest stratum must be the one that was originally at the bottom, and the strata above it are progressively younger, with the youngest on top. This is now known as the *law* (or principle) *of superposition of strata.* He used this line of reasoning to show that Tuscany had twice been flooded by the sea and then, following the retreat of the water and later erosion, it had formed level plains. Mineral veins were different. These must have flowed as liquid through rocks that were already hard, and crystals formed as the liquid cooled or remained stationary for a long time.

Steno had advanced a convincing explanation for the way fossils form, but he had done much more than that. He had provided scientists with a way to trace the history of the rocks around them. This branch of Earth science is called *stratigraphy,* and Steno is often called the father of stratigraphy.

Stratigraphic dates are relative, not absolute. Although rock strata form sequentially, adjacent strata might have formed millions of years apart or only weeks apart. Steno offered his own interpretation. He believed that sedimentary rocks could be dated in respect of the biblical flood. He had observed that a layer of rocks in the Appenine Mountains contained no fossils, but a layer above it contained many. He concluded that the lower layer formed before life was created and that the upper layer formed during the flood, when life existed. In his studies of Tuscan strata, he suggested that the earlier flood occurred on the second day of creation, when an ocean covered the entire world, and the second flood was the one described in Genesis. He suggested that the second flood occurred 1,650 years after the first flood, and that the world was created in 4000 B.C.E.

FOSSIL FUELS

Let me tell you next of stones that burn like logs. It is a fact that throughout the province of Cathay there is a sort of black stone, which is dug out of veins in the hillsides and burns like logs. These stones keep a fire going better than wood. I assure you that, if you put them on the fire in the evening and see that they are well alight, they will continue to burn all night, so that you will find them still glowing in the morning.

That is how Marco Polo (1254–1324) described coal when he first encountered it in Cathay (northern China) toward the end of the 13th century. The story began in 1260 when two Venetian merchants, the brothers Niccolò and Maffeo Polo, sailed from Constantinople (Istanbul) to Sudak on the Crimean coast (in modern-day Ukraine), where a third brother, Marco, owned a house. They were seeking better markets for their wares, and Maffeo and Niccolò moved on to the town of Surai on the banks of the Volga River. War broke out after they had been trading there for a year, and in trying to find a way back to Italy they became stranded in Bukhara, in Uzbekistan. An emissary from the Mongol Empire to the east rescued the brothers, persuading them to accompany him to the court of Kublai Khan. The Great Khan, he said, had never seen any Latins (Europeans) and would be pleased to meet them. So they traveled to Beijing, where the khan had established his court. After a year, Kublai Khan sent them back to Europe with a message for the pope, requesting 100 learned men to teach him about Christianity and Western science. In 1269 the brothers arrived at Acre, at the eastern edge of Christendom, and from there returned to Italy. Pope Clement IV had died in 1268, and for three years there was no pope, so in 1271, having failed to obtain any response to the khan's request, the brothers departed again for Acre, fearing they would lose their markets if they delayed longer. At Acre they enlisted the support of the papal legate Tedaldo Visconti and had just departed for Palestine when they were recalled. The new pope, Gregory X, had blessed their journey and gave them full diplomatic credentials. He was unable to send 100 men to Beijing, but he did offer two. These individuals lacked the courage and resourcefulness of the Polos, however, and the brothers set off without them, this time taking with them Niccolò's son Marco, a boy of 17. The three men spent the next 24 years in Asia. Marco's account of their adventures, *The Travels of Marco Polo,* dictated to a fellow captive, Rustichello of Pisa, while he was a prisoner of war in Genoa, is one of the most famous travel books in the world.

Clearly his first sight of coal surprised Marco, but its use was far from new. The Chinese had been burning coal for heating and for smelting ores since the Warring States Period (475–221 B.C.E.). The discovery of large quantities of coal fragments in the ditches surrounding a Late Iron Age settlement near present-day Edinburgh, Scotland, shows that the British were burning coal more than 2,000 years ago. Still earlier Aristotle referred to a rock resembling charcoal.

Centuries later coal became the world's principal fuel, and mining it grew into a vast industry. That industry is still expanding in some countries, especially China, but in others it is in decline. The photograph shows miners at La Houve Mine, in Creutzwald, France, on April 23, 2004, on the day it closed after 300 years of operation.

Marco Polo may well have wondered at this curious black rock. No other rock will burn, giving off as much heat as charcoal, but it was not until the opening years of the 19th century that scientists discovered why coal possesses this property. Naturalists specializing in mining already understood that the Earth has a long history. Abraham Gottlob Werner (1749–1817), the highly influential professor of mineral science at the Mining Academy in Freiberg, Germany (see "Abraham Gottlob Werner—and the Classification of Rocks," pages 114–117), taught that it was possible to use the study of rocks and minerals to reconstruct the history of the Earth, but Werner paid little or no attention to fossils and the contribution they might make to the unravelling of that history. Others did, however, most notably the French zoologist Georges Cuvier (1769–1832), professor

Coal miners bring out the last blocks of coal from La Houve Mine, in Creutzwald, on April 23, 2004. La Houve was France's last working coal mine, and on that day it closed. *(Jean-Christophe Verhaegen/Stringer)*

of comparative anatomy at the Natural History Museum in Paris, who founded modern paleontology. Cuvier was a zoologist, studying fossil bones, and it was Johann Friedrich Blumenbach (1752–1840), a professor of medicine at the University of Göttingen, in Germany, who attempted to place the fossil bones into a historical context.

Another Göttingen scholar, Ernst Friedrich von Schlotheim (1764–1832), had studied law but also attended Blumenbach's lectures on natural history. He went on to the Mining Academy at Freiberg and, his formal education completed, became a civil servant in Gotha, the German state where he had been born. He retained his interest in Earth sciences, and especially in fossil plants. In 1804 Schlotheim published *Plant Fossils*, a fully illustrated work on what its author described as "the flora of the former world." The engraved illustrations included fossil impressions of leaves, stems, and other fragments of plants from coal mined in Thuringia, Germany. Schlotheim was unable to identify these plants by comparing them with pictures of living plants, but the coal measures from which they came were known to lie lower than certain other *sedimentary rock* strata. They must therefore be much older than the rocks above them. A picture was emerging of a sequence of historical ages, each with characteristic plants and animals preserved as fossils. Coal, then, was identified as the remains of vegetation from a time in the distant past, a time, Schlotheim wrote, when Germany was a region of swamp forests, palm trees, tree ferns, and giant horsetails.

Petroleum and natural gas are also the remains of once-living organisms, in this case marine organisms. Coal formation begins when plants fall into mud or shallow water, where the lack of oxygen prevents their complete decomposition. The remains are then compressed and heated. Petroleum forms when organic material is buried beneath seabed sediments that completely isolate them and is then subjected to very high pressures and temperatures. The material is cooked until it turns into a liquid that permeates the porous structure of a rock mass. It contains no fossils, so there is no obvious way to recognize its origin or date its formation. Ignorance of its origin did not prevent its usefulness from being recognized.

There are certain places where petroleum seeps out at the ground surface and uses were found for it a very long time ago. More than 5,000 years ago people in Mesopotamia were using asphalt—a dark-colored, viscous liquid that sets hard on exposure to air—to fix handles to the blades of tools, to caulk boats, and to set jewels and mosaics. It was sometimes used as a mortar in building, as in Babylon, and to surface roads, making them more suitable for wheeled traffic. By the time Marco Polo reached the shores of the Caspian Sea in Georgia, early in his travels, oil was being traded. He reported: "Near the Georgian border there is a spring from which gushes a stream of oil, in such abundance that a hundred ships may load there at once. This oil is not good to eat; but it is good for burning and as a

(continues on page 112)

A jack pump, or "nodding donkey," pumping oil from a well. Once a familiar sight in oil mining areas, new extraction technologies are rendering these pumps obsolete. *(Martin Bond/Photo Researchers, Inc.)*

THE DISCOVERY OF METHANE HYDRATES

In the 1930s oil and gas engineers began to experience trouble with the blocking of pipelines carrying oil from fields in the far north. They found the culprit was a substance that until then had been nothing more than a chemical curiosity: clathrate hydrates.

Crystalline solids form lattices, which are structures in which the individual atoms are arranged in a very regular pattern. A clathrate is a structure in which small molecules of one substance are held in the spaces inside the lattice of another substance. In the case of a clathrate hydrate, that other substance is ice. The pipelines were choking on ice that contained several substances derived from petroleum. The clathrate hydrates, resembling ordinary ice but much stronger, clung to the sides of the pipes, forming a layer that quickly thickened. No sooner had the engineers cleared a pipe than it began to fill once more.

The hydrates form when water freezes under high pressure. They are found below the surface in seabed sediments and *permafrost*. Permafrost is soil in which the water remains permanently frozen. Under these conditions the ice can freeze into a 12-sided, three-dimensional shape called a dodecahedron. Each of its faces is a pentagon, and the companion gas is held inside the hollow "ball."

Clathrate hydrates were a nuisance and added to the cost of pumping oil, but by the 1980s they were beginning to arouse a different interest. At first it was assumed that they occur only in the Arctic and around its borders, but as the oil industry continued to expand and offshore oil fields were opened all over the world, this proved to be wrong. Oil-industry scientists found there were clathrate hydrates below the seabed almost everywhere they looked. Only a very small proportion of them are held in permafrost on land. The map shows the approximate location of known and suspected sources of them.

The most abundant gas associated with them is methane, so they are now known as methane (or gas or natural gas) hydrates. Methane is the principal constituent of natural gas, which is one of the most important fossil fuels. Scientists estimate that in the world as a whole there are between 35 and 61,000 quadrillion (35–61,100 $\times 10^{15}$) cubic feet (1–1,726 $\times 10^{15}$ m^3) of methane held in hydrated form in sediments within 6,500 feet (2,000 m) of the surface. That is more than double the total of all the other fossil fuel reserves in the world, and its location beneath relatively shallow coastal waters places it potentially within reach of the offshore oil industry. At present it is uncertain whether the hydrates could be mined economically for their methane, but if ways are found to obtain the methane, it will be by far the world's most important source of fuel. Methane is also the most benign of all the fossil fuels, because it burns cleanly and releases less carbon dioxide than oil or coal for every unit of energy.

Their widespread distribution has two implications, apart from the difficulties the ices cause for oil and gas pipelines.

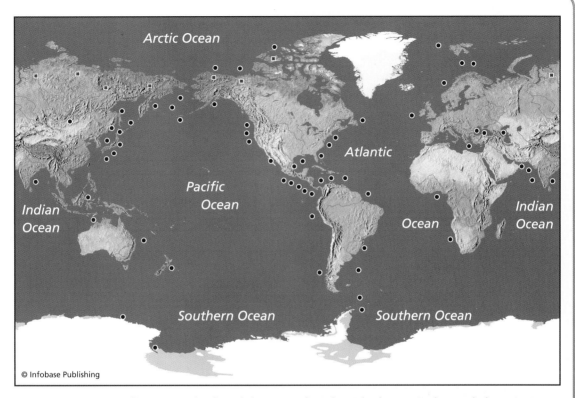

Worldwide location of known and inferred deposits of methane hydrates. Circles mark deposits in sediments on continental shelves; squares mark deposits in permafrost areas on land.

The first worry relates to what might happen if the methane were to be released accidentally. Some scientists believe that sudden large releases can sink ships quickly and with no warning. This is because a ship floats because, with the air it contains, it is less dense than seawater. When a huge amount of gas bubbles up from the seabed, however, what a moment ago was water becomes a mixture of water and gas, and the density decreases dramatically. It happens suddenly, and the effect lasts no more than a minute or two, but that may be long enough to doom the ship by removing its support. While the effect lasts, the ship is much denser than the water, so it sinks vertically downward like a stone and fills with water.

The second possibility is that a massive release of methane might contribute to climate change. Methane is a greenhouse gas 10 times more effective than carbon dioxide at absorbing heat.

Besides the fears, though, there is an exciting possibility.

(continued from page 109)

salve for men and camels affected with itch or scab. Men come from a long distance to fetch this oil, and in all the neighborhood no other oil is burnt but this." In North America, Native Americans had also discovered the uses of oil, which they would skim from the surface of water near natural seeps. They used it as a medicine and to make their canoes waterproof. The Chinese came across oil when they were mining salt and were the first people to drill oil wells, in the fourth century C.E. or possibly earlier. They burned the oil to evaporate brine for the production of salt.

Oil mining and refining continued on a small scale until the 20th century and the introduction of the internal combustion engine. Then demand grew steadily, and "nodding donkeys" of the type shown in the photograph became a familiar sight in those parts of the world where oil was being pumped from below ground.

Petroleum geologists now understand how petroleum formed and how natural gas, principally methane (CH_4), is one of its chemical constituents. This knowledge was acquired during the latter part of the 20th century, when the oil industry had grown to a large size and the commercial value of oil justified high investment in scientific research. For most of history, however, oil's origin was a mystery.

There are fears that a time may come, perhaps soon, when oil production will peak and prices will rise sharply. Oil and gas can be made from coal, but there may be an alternative: methane hydrates (see sidebar). Discovered in the 1930s, methane in this form has now been found in many parts of the world. At present it is not being extracted, but there is intensive research into the technologies needed to gain access to it. If these succeed, the world may possess much more methane than all its other reserves of fossil fuels combined.

Age of the Earth

Several billion years ago a cloud of gas and dust contracted by gravitational attraction began to rotate, and pressure due to the accumulation of mass at its center ignited a thermonuclear reaction that made the center shine. It had become a new star, the Sun, and about 4.6 billion years ago the solar system was born. Earth is the third planet counting outward from the Sun, and it was made from gas, countless millions of dust particles, and rocks that collided and stayed together. Soon after the protoplanet had formed, it collided with a rock the size of a small planet. Both bodies were shattered to fragments, and when the fragments reassembled, they formed two bodies: Earth and the Moon.

It is impossible to imagine what 4.6 billion years really means. It is just a big number, so much bigger than the lifespan of a person or, indeed, of the whole of human history that the mind cannot truly comprehend it. Throughout most of history people had no way to measure the age of the Earth or to unravel details of the events that formed and shaped it, but by the middle of the 18th century scientists were coming to realize that the planet is extremely ancient. In the 20th century they learned how to date rocks by the very steady decay of radioactive elements contained within them, and from that they were able to calculate the age of the Earth itself.

This chapter tells of the observations that demanded explanations, and the explanations that led to the unravelling of the Earth's long history. History is made up of a sequence of events: It implies change, and the chapter begins with the ideas of Abraham Gottlob Werner, the most famous geologist of his day. Those ideas were challenged by one of his students, Alexander von Humboldt, the Prussian geographer and Earth scientist who was among the first to note that the Earth changes, that it has a history. Ice ages are among the most dramatic changes, and the story describes their discovery. Whereas the previous chapter outlined the recognition of the fact that fossils are the remains or traces of once-living organisms, this chapter tells of their use for dating rock strata. Various explanations were offered for the origin of rocks and the cause of the events they record. Once thought to have been settled, a major controversy between uniformitarianism and *catastrophism* reignited in the 1970s with the discovery of evidence for a catastrophic event 65 million years ago that led to the extinction of the dinosaurs.

Finally scientists were able to compile a coherent summary of Earth history. The geologic timescale supplies names and dates for all the episodes from the formation of the Earth to the present day. The chapter ends with the present version of that timescale on pages 151–152.

ABRAHAM GOTTLOB WERNER— AND THE CLASSIFICATION OF ROCKS

Earth scientists discussing the history of the planet sometimes refer to the Tertiary and Quaternary suberas. "Primary," "secondary," "tertiary," and "quaternary" divide the history of the Earth into four episodes, an idea that grew from one proposed in Germany in the 18th century by a number of mineralogists, most notably Johann Gottlob Lehmann (1719–67), a mining engineer.

Mining for metals was of great economic importance throughout the independent states that would later unite to become Germany, and there was intense interest in the way minerals form. The underlying theory of mineralization had been developed from the work of Agricola (see "Agricola—and the Formation of Ores," pages 87–89). The process was believed to be entirely chemical. Just as crystals grow in a strong solution of salt or sugar as the water evaporates, so German mineralogists believed that all mineral crystals form from aqueous solutions as the water evaporates. Crustal rocks are rich in granite, which is crystalline. This suggested that at one time oceans had covered the entire Earth. Scientists observed that the Baltic Sea, bordering Germany, is growing steadily shallower. Scientists now know this is because the seabed is rising following the removal of the weight of ice sheets at the end of the last ice age (see "Ice Ages," pages 122–126). Ice ages had not been discovered in the 18th century, however, and the changes in the Baltic Sea were then seen as evidence that the oceans were steadily retreating.

Lehmann developed this concept in his 1756 book *Versuch einer Geschichte von Flötz-Gebirgen, betrefend deren Entstehung, Lage, darinne befindliche Metallen, Mineralien und Fossilien* (Attempt at a history of sedimentary rocks, with regard to their origin, position, and their metal, mineral, and fossil content), dividing the rocks of the Earth's crust into three classes. Originally, Lehmann argued, a vast ocean had covered the Earth. Minerals crystallized from that ocean to form the rocks at the core of mountain ranges. These he called primary rocks; they contained the most valuable minerals and ores. Some time later the Earth was inundated once again during the biblical flood, and sediments covered the mountains, solidifying to form secondary rocks. When the floodwaters receded, later erosion and sedimentation produced tertiary rocks. When interpreted historically, this division gave the names "primary," "secondary," and "tertiary" to periods of time.

The most influential German school of mining was at Freiberg (now the Technische Universität Bergakademie Freiberg) in Saxony, founded in 1765. In 1775 a new principal, Abraham Gottlob Werner, arrived to take charge. Werner continued teaching at Freiberg for the rest of his life and became the most famous teacher of mining and geology in Europe. Students flocked to Freiberg from far and wide to study under him, returning to their homes to spread his teachings. Werner agreed with the broad outline of Lehmann's idea but took it further, and because of his reputation as a teacher, it became widely accepted.

Werner was born on September 25, 1750, near Bunzlau, Silesia (now Bolesławiec, Poland). The family had been involved in mining engineering for several generations, and Werner's father was an inspector at an ironworks. Abraham studied at the Freiberg Mining Academy from 1769 until 1771, completing his education at the University of Leipzig.

Werner developed Lehmann's ideas basing his own theories on a detailed and precise identification of minerals, but he recognized that the history of the rocks was more complicated than Lehmann might have supposed. In Werner's history the complications arose because of the irregular shape of the Earth's surface. The primary rocks crystallized out of the world ocean directly onto the solid surface, following the surface contours. At first, therefore, the entire surface was covered by granite and basalt. Primary rocks contain no fossils. As the ocean retreated, the more elevated parts of the mountains were exposed to the air. The exposed rocks eroded, producing the sediment that formed successive layers of secondary and tertiary rocks, identifiable from their distinctive suites of minerals. These rocks do contain fossils. Once the sedimentary rocks had formed, they settled and moved a little, causing folding and faulting.

According to the Wernerian theory all rocks form below sea level, consequently the height of the sea determines the elevation of rocks. Younger rocks are deposited above older rocks, and because the oceans have been retreating throughout history, rocks become progressively older with increasing height above sea level. The oldest rocks are to be found at the summits of mountains.

As he acquired more evidence, Werner modified his theory. He recognized that many rock formations contain minerals he had thought typical of different periods of deposition. He also came to accept that basalt had solidified from a molten state and was of volcanic origin. Werner did not think volcanism important, however. He believed that it occurred only locally and that the rocks it formed were recent. To those who suggested granite had solidified from molten rock, Werner pointed out (correctly) that its structure is quite unlike that of solidified lava. Werner also attempted to date rocks by the fossils they contain, suggesting that organisms became more complex as successive phases of rock formation purified the waters of the oceans.

Werner based his theories on his intimate knowledge of the geology of Saxony, assuming that rocks throughout the world were similar to those of Saxony. The principal difficulty with his theory was its complete lack of any explanation of where all the water came from and disappeared to. Challenged with this, Werner said the evidence was in the rocks, and all he did was interpret what he observed. The origin and fate of the water was not his concern. He shared the widespread assumption that the Earth was completely rigid. It was only later that evidence accumulated to show that rocks can subside and be uplifted, altering their elevation with respect to sea level. Werner died in Dresden, on June 30, 1817.

ALEXANDER VON HUMBOLDT, WHO RECOGNIZED THAT THE EARTH CHANGES OVER TIME

Many of Werner's students achieved eminence in later life, but none became so renowned as Alexander von Humboldt (1769–1859). One of the most influential thinkers of the 19th century, Humboldt, to a large extent, shaped the scientific study of the natural environment. He became what today would be called a celebrity. He was said to have been the second most famous person in Europe, after Napoléon Bonaparte, and when he died, he was given a state funeral.

Humboldt was born in Berlin on September 14, 1769, an aristocrat who inherited the family title (baron) on the death of his elder brother. His father was closely involved in the political and intellectual life of Berlin, the capital of Prussia, which was the most important and powerful of the German states, ruled by Friedrich II (1712–86), known to the world as Frederick the Great.

Alexander's father died in 1779, the year that Humboldt's education commenced. Until 1789 he and his brother Wilhelm (1767–1835) studied under private tutors, then they both enrolled at the University of Göttingen, where Alexander became very interested in natural history and especially in minerals. He met the distinguished German naturalist Georg Forster (1754–94), who had sailed with the English explorer Captain James Cook (1728–79) on Cook's second voyage around the world, and in 1790 Humboldt and Forster traveled Europe together, visiting several eminent scientists of the day. Humboldt returned to Prussia enthused by science and determined to obtain a sound scientific education. In 1791 he enrolled at the Freiberg Mining Academy, studying under Werner and graduating after two years. Afterward Humboldt spent the years 1792 to 1797 on a diplomatic mission touring the salt-mining regions of Europe, during which, in 1793, he was appointed assessor and later director of mines in the region around Bayreuth.

Humboldt's life changed in 1796 when his mother died and he inherited a share of the family's considerable fortune. He no longer needed to earn a living and was free to indulge his passion for travel.

With his friend, the French botanist Aimé Bonpland (1773–1858), in 1799 he embarked on a journey to the Spanish-American colonies. The two explorers discovered that the headwaters of the continent's two greatest rivers, the Amazon and Orinoco, are linked. They collected samples from more than 60,000 plants, approximately 6,000 of which were previously unknown to botanists, and Humboldt noted the way vegetation changes with elevation and latitude, making him one of the founders of biogeography. Humboldt discovered the Peru Current, also known as the Humboldt Current, and he and Bonpland climbed higher than any mountaineer before them, reaching 18,893 feet (5,762 m) on Mount Chimborazo, in Ecuador. Humboldt was the first person to identify lack of oxygen as the cause of the mountain sickness he and his friend suffered. He studied, measured, and recorded variations in the geomagnetic field, advanced the science of meteorology, and sent samples of guano back to France for analysis, indirectly starting the industry exporting guano for use as a fertilizer. As he climbed through the South American Andes, Humboldt became increasingly interested in volcanoes and the extent of volcanic activity. He noted that in Ecuador many of the volcanoes form a straight line, concluding that this is because they sit above a fault in the rocks deep below ground. Most important, he observed that many of the rocks he saw were of volcanic origin.

Alexander von Humboldt in 1813, when he was 44 years old and living in Paris. The engraving is by A. Krausse. *(Science Photo Library)*

In 1779, after his return to Europe, Humboldt, his friend the French chemist Jean-Louis Gay-Lussac (1778–1850), and the German geologist and geographer Christian Leopold von Buch (1774–1853) visited Mount Vesuvius, which erupted while they were there. Buch had studied at Freiberg and, like Humboldt and most of Werner's students, had accepted the Wernerian idea of an aqueous origin for all crystalline rocks. But the evidence of the extent of volcanism he had seen in Europe and Humboldt had seen in the Andes convinced both men that Werner was mistaken. Crystalline rocks result mainly from volcanic eruptions and the cooling and solidification of molten rock, and volcanism has repeatedly altered the surface of the Earth. The illustration, from an engraving by A. Krausse, shows Humboldt in 1813. By 1827 his fortune was spent, however, and when the Prussian king offered him a paid position at court Humboldt was in no position to refuse. He returned to Berlin in 1827, as chamberlain to the court. Humboldt died in Berlin on May 6, 1859.

COMTE DE BUFFON—AND THE COOLING EARTH

Werner's belief in a world once covered by oceans revived the theory of *Neptunism* (see "Neptunism," pages 142–144), which aimed to explain how rocks form but beyond recognizing that the process must be slow, had nothing to say about the age of the Earth or the manner of the planet's formation. One of the first proponents of Neptunism did propose an age for the Earth, however, and also a mechanism by which it might have formed.

Georges-Louis Leclerc (1707–88) was a naturalist, but one with wide interests, and he was also a prolific and gifted writer, who achieved fame and fortune and received many honors. He was born into a wealthy family on September 7, 1707, in Montbard, Burgundy, in France. Between 1714 and 1717 his parents bought a large estate that included the land around Buffon and added *de Buffon* to their name. Leclerc attended the Jesuit college at Dijon, graduating in law in 1726, but while a student at Dijon, he developed a keen interest in mathematics and astronomy. In 1728 he moved to Angers, where he studied medicine, botany, and probably mathematics. His mother, Anne-Christine Marlin, died in 1731, and the large legacy he inherited from her enabled Leclerc to travel and pursue his scientific interests. In 1732 he moved to Paris, where he lodged with the apothecary to the king, and in 1733 he was elected an associate member of the Royal Academy of Sciences. His study of the properties of timber, using trees on his Burgundy estate, led to his appointment in 1739 as keeper of the Royal Botanical Garden. He was elected a fellow of the Royal Society of London in 1738 and a member of the Académie Française in 1753, where his inaugural address was his *Discours sur le style* (Discourse on style), which became very famous. In 1771 he was made a count, henceforth becoming the comte de Buffon. He died at the Royal Botanical Garden in Paris on April 16, 1788, and is buried in Montbard.

Buffon's major work was the highly popular *Histoire naturelle génerale et particulière (Natural History, General and Particular)*, an encyclopedic work covering the whole of natural history that ran to 44 volumes, published between 1749 and 1804. The work underwent several editions and was translated into German, English, and Dutch. In it Buffon introduced some highly original ideas on the history of the Earth.

The first of these, called "History and Theory of the Earth," appeared in 1749 in the first volume of the *Histoire naturelle*, where Buffon intended it to establish the physical nature of the planet whose animals and plants the later volumes would describe. He maintained that the history of the Earth must be described in terms of constant, continual, widespread processes that can be seen operating today. Fossil seashells are found in rocks far from the sea, and Buffon took this to imply that at different times oceans had covered all of the Earth. Currents and tides sculpted submarine mountains and other features, and sedimentary rocks were deposited on the ocean floor. Surfaces on dry land eroded and were carried away by rivers, to supply fresh sediment to the oceans. Any point on the Earth's surface might have been both dry land and ocean floor at different times. It was a directionless process of continual, gradual change that had always been and would continue to the end of time. This was his contribution to the theory of Neptunism.

Although Buffon had explained the history of Earth, he had said nothing about the way the world might have come into existence. He completed the story in 1778 in a supplementary volume, *Les époques de la nature* (Nature's epochs), of which he said there were seven. During the first epoch the solar system began when a comet collided with the Sun. In the 18th century scientists believed comets were dense, very solid bodies rather than the loose piles of ice, rubble, and dust often called "dirty snowballs" they are now known to be, so a collision with a comet would have been a major event. The comet struck the Sun at an oblique angle, dislodging a plume of solar matter that contracted as it cooled, breaking into globules that condensed into the planets. The theory explained why all the planets lie approximately in the same plane and orbit the Sun in the same direction. Each planet cooled at a different rate depending on its size and, presumably, its distance from the Sun. Buffon estimated that it would have taken the Earth at least 70,000 years to reach its present temperature. As the Earth cooled, the crust solidified into granite. As soon as the crust was below the boiling temperature of water, boiling rain began to fall, eventually covering the entire planet in a hot ocean. The water eroded the recently cooled rocks, producing sediment that formed layers of sedimentary rock that were laid down in subsequent epochs. The granite mountains are all that remains today of the original crust.

The presence of fossils in the sedimentary beds demonstrates that animals inhabited the early ocean, but Buffon knew that many of these fossils were of animals that no longer exist. He explained this by saying that the ocean's first inhabitants were able to live in water that was almost boiling, and they died as the temperature fell below what they found tolerable, to be succeeded by others adapted to different, cooler temperatures. Dry land also cooled slowly and the presence of the fossils of giant animals such as mammoths in northern North America and Siberia showed that these regions had once been much warmer than they are today and that as they grew colder their inhabitants migrated toward the equator, where their descendants, the elephants and hippopotamuses, live today. In volume 5 of *Les époques* Buffon wrote the following:

> ... the earliest and greatest formation of animated beings occurred in the high, elevated regions of the north, from whence they have successively passed into the equatorial regions under the same form, without having lost anything but their great size; our elephants and hippopotamuses, which appear large to us, had much larger ancestors during the time in which they inhabited the northern regions where they have left their remains.

Water, Buffon believed, is able to penetrate into the Earth. There it engages in chemical reactions that generate steam, and the expanding steam causes volcanic eruptions, which he suggested began fairly recently in Earth's history. It explains why all volcanoes are found near the coast.

Buffon's accounts were entirely plausible, given the amount of hard information available to scientists of his day, but they contradict each other. Buffon began by insisting that Earth is in a state of dynamic equilibrium, constantly changing while remaining overall the same. Yet his account of Earth's origin is cataclysmic, a unique and extremely violent event of the kind he had already said did not happen. That contradiction continued to raise controversies, often heated ones, until a compromise was reached in the second half of the 20th century. The need for compromise became evident with the discovery of past ice ages.

ICE AGES

Buffon's idea that mammoths and mastodons lived in a much warmer climate and that elephants are their direct descendants did not survive for very long. Georges Cuvier (1769–1832), the most eminent zoologist of his day and the founding father of paleontology, showed that mammoths and mastodons were different from elephants and might have lived in temperate or even cold climates. Their fossils could not be taken as evidence that northern regions were once warmer than they are now. All the same, fossils of tropical plants had been found in high latitudes and the high temperatures encountered in deep mines strongly suggested that the interior of the Earth is warmer than its surface. The French mathematical physicist Jean-Baptiste-Joseph Fourier (1768–1830), who specialized in the mathematics of heat transfer, maintained that the heat at the Earth's center was best explained if the entire Earth had once been hot and it was cooling from the outside inward. So, Buffon's idea of a cooling Earth had strong support. In the latter part of the 19th century its most outspoken supporter was William Thomson, Lord Kelvin (1824–1907).

The wide acceptance of the idea of a gradually cooling Earth led to a difficulty, however. All over northern North America and Europe there are beds of gravel and large rocks that are of entirely different types to the solid bedrock beneath them. Such deposits and boulders are called *erratics,* and they were a mystery. How did they come to be there? Most geologists believed they had been hurled long distances by huge *tsunamis* caused by mighty upheavals in the crust that sent floodwaters far inland.

In 1815 Jean-Pierre Perraudin (1767–1858), who spent much of his time hiking and hunting chamois in the Swiss Alps, wondered how the large granite boulders he observed high on hilltops had come to be there. Surely, he thought, no flood could have lifted them so high. He also noticed long scratches on rocks lining mountain valleys. Perraudin thought *glaciers* might have pushed the boulders along,

An alpine glacier near the French-Italian border, showing the alternating light and dark bands made as the ice moves downhill. In summer, rocks scoured from the valley sides make the dark bands; in winter, the movement is slower, with less scouring resulting in white bands. *(Ted Kinsman/ Photo Researchers, Inc.)*

scratching the adjacent rocks as they did so, but that implied that the glaciers once extended much farther down the valleys. The illustration of an alpine glacier shows how plausible this idea is. Perraudin approached Jean de Charpentier (1786–1855), an established Swiss geologist, with his idea. Charpentier rejected it, but another naturalist, Ignace Venetz (1788–1859), was persuaded. Venetz delivered a lecture in 1821 outlining the features of the Swiss landscape that could be explained if glaciers were able to flow and if they were once more extensive. A later lecture in 1829 finally convinced Charpentier, who had turned his attention to the geology of the Swiss mountains following a disaster in 1818, when an ice dam broke and caused many deaths by drowning.

Charpentier was an enthusiastic convert and began his own investigations, but he was opposed by the most famous Swiss scientist of

AGASSIZ, CHARPENTIER, AND CROLL

Jean Louis Rodolphe Agassiz was born on May 28, 1807, at Motier, near Lake Morat, Switzerland. His father was a Protestant pastor, and his mother, Rose Mayor, taught him to love the natural world. He was educated at the gymnasium (high school) at Bienne (Biel), near Bern, and later at a school in Lausanne. In 1824 he enrolled at the University of Zurich and in 1826 moved to the University of Heidelberg, Germany, but caught typhoid fever and had to return to Switzerland. In 1827 he enrolled at the University of Munich, Germany. He qualified as a doctor of philosophy at the University of Erlangen in 1829 and as a doctor of medicine at Munich in 1830.

Agassiz specialized in the study of fossil fishes, and in November 1831 this took him to the Natural History Museum in Paris, where he worked under the supervision of Georges Cuvier (1769–1832). Cuvier and Agassiz became friends, and after Cuvier's death Alexander von Humboldt (1769–

1859) helped Agassiz obtain the professorship of natural history at the University of Neuchâtel, Switzerland. He continued to classify fossil fish, but in 1836 he turned his attention to a different problem, of investigating the boulders and gravel deposits, known as erratics, that lay scattered across the plain of eastern France and were unrelated to the underlying rocks. He studied the Aar Glacier over several years and discovered conclusive evidence that glaciers flow like extremely slow rivers. He became convinced that in the geologically recent past all of Switzerland and those parts of Europe where erratics were found had been covered by an ice sheet similar to that covering most of Greenland.

In 1846, with the help of a grant from King Friedrich Wilhelm IV of Prussia, Agassiz visited the United States, partly to continue his studies and partly to deliver a series of lectures at the Lowell Institute in Boston. He followed these the day, Jean Louis Rodolphe Agassiz (1807–73). Charpentier eventually persuaded Agassiz to make a field trip into the Alps to examine the evidence, and in the summer of 1836 they visited the Aar Glacier. Agassiz continued to study the Aar Glacier for several more years (see sidebar). In 1839 he noted that a hut, which had been erected on the ice in 1827, had moved one mile (1.6 km) from its original position. He drove a straight line of stakes across the glacier from side to side and found that two years later the line was no longer straight; it was U-shaped because the stakes at the center had moved farther than those at the sides.

Fully convinced, on July 24, 1837, Agassiz devoted his presidential address to the Société Helvétique de Sciences Naturelles in Neuchâtel to the proposition that at one time ice sheets had covered the Northern Hemisphere from the Arctic all the way to the Mediterranean.

with lectures in other cities and extended his stay. In 1848 he was appointed professor of zoology at Harvard University. He became an American citizen and remained in the United States, mainly at Harvard, for the rest of his life. He found evidence that North America had also been covered by ice, and he traced the shoreline of a vast, vanished lake, now known as Lake Agassiz, that once covered North Dakota, Minnesota, and Manitoba. He developed the Museum of Comparative Zoology at Harvard and became one of the country's finest teachers of science. Agassiz died in Cambridge, Massachusetts, on December 12, 1873. A boulder from the Aar glacial moraine marks his grave.

Jean de Charpentier was born on December 8, 1786 in Freiberg, Saxony, Germany, where his father was a professor at the Mining Academy. Charpentier studied at the academy under Abraham Gottlob Werner, and after graduating he worked as an engineer in the salt mines at Bex,

Switzerland. He died at Bex on December 12, 1855. In the early 1830s Charpentier assembled observations of moraines, striations in rocks, and erratics, together with the position and form of existing Swiss glaciers, and from this proposed that glaciers had once been far more extensive.

James Croll was a Scottish climate scientist and geologist, born at Cargill, Perthshire, on January 2, 1821. His family could not afford to pay for his education, and he left school at 13. Croll worked at a succession of jobs in order to earn money but was an avid reader, educating himself in his free time. In 1867 Croll was placed in charge of the Edinburgh office of the Geological Survey of Scotland, a post he retained until his retirement in 1880. He died from heart disease on December 15, 1890. In 1864 Croll proposed that the onset of glacial periods (ice ages) is triggered by changes in the eccentricity of the Earth's solar orbit and the precession of the equinoxes.

Temperatures had fallen so fast that all life had been destroyed. Charpentier, though no less convinced, had reached a more cautious conclusion, maintaining only that the Swiss glaciers had extended to lower levels following a prolonged period of cold weather. Agassiz's view prevailed, however, because he was first to publish. His *Études sur les glaciers* (Studies of glaciers) appeared in 1840.

Agassiz's lecture was received with a mixture of shock and rage. The idea that temperatures in the past had been lower than those of the present was wholly irreconcilable with the prevailing idea of gradual cooling, and the glacial theory contradicted the idea that the Earth had been flooded. Agassiz accepted that the Earth had cooled gradually but maintained the cooling had not happened evenly; at times it had accelerated then slowed again.

Little by little the scientists were won over. Humboldt remained unconvinced, Charles Darwin (1809–82) was partly persuaded, but most geologists took much longer to accept the idea. Charpentier's ploy of persuading a doubting Agassiz to visit a glacier and examine the evidence helped. Agassiz persuaded the highly influential English geologist William Buckland (1784–1856) to visit the Alps in 1838, and although Buckland believed firmly in the biblical account of the Flood, he could not deny the evidence before him. Buckland accepted Agassiz's glacial theory and promoted it in Britain. In 1840 Agassiz visited Scotland, where he found further evidence of past glacial activity, and in 1846 he visited the United States, where he also found evidence of glaciation, supporting his concept of a "Great Ice Age" that in comparatively recent times had gripped the entire Northern Hemisphere.

LAYERS OF ROCK

Werner taught a generation of mining engineers and geologists that rocks form under water (see "Abraham Gottlob Werner—and the Classification of Rocks," on pages 114–117). More important, he taught that they form clearly defined layers. Miners and quarry workers were very familiar with the rock layers they exposed as they cut vertically downward into the Earth, and two fundamental principles of rock layers had been enunciated almost a century earlier by Steno (see "Nicolaus Steno, Who Fully Understood Fossils," pages 102–105). Steno asserted (correctly) that sedimentary rock layers were originally horizontal. This is known as the law of original horizontality. Steno also asserted that younger rock layers lie on top of older layers; this is the law of superposition of strata.

During the late 17th and early 18th centuries miners and naturalists used these two laws to arrange rock strata in their chronological order. Werner called his study of the rocks "geognosy" (sometimes it was called "geognosis") and its practitioners "geognosts." Since about 1800 the preferred terms are *geology* and *geologists,* respectively, but for a time both were in use. Geognosts then began naming the strata, and building what is now known as the *stratigraphic column.*

In 1760 Giovanni Arduino (1714–95), a Venetian mining engineer, began the process. He gave the name *Primary* to the mica slates found in the Atesine Alps in the north, *Secondary* to the mountain limestones with fossils of marine organisms found in the foothills of the Alps, *Tertiary* to the fossil-rich sedimentary rocks of the valleys, and *Quaternary* to the rocks of the floodplain of the Po River. These were the names Werner adopted later. In 1795 Humboldt (see "Alexander von Humboldt, Who Recognized That the Earth Changes over Time," pages 117–119) gave the name *Jurassic* to rocks he found in the Jura Mountains of Switzerland. In 1822 the English geologists William Daniel Conybeare (1787–1857) and William Phillips (1775–1828) published *Outline of the Geology of England and Wales,* in which they named the *Carboniferous* strata they found in the north of England. In the same year the Belgian geologist Jean-Baptiste-Julien d'Omalius d'Halloy (1783–1875) named the *Cretaceous* rocks of France.

Phillips, the son of James Phillips, a London printer and bookseller, also wrote a popular book on geology, *Outlines of Mineralogy and Geology,* published in 1815. In it Phillips, a founding member of the Geological Society, described all the rock formations in order, from granite to alluvium, emphasizing that the scientific progress then being made so rapidly was due to the concentration on "facts" rather than "theorizing.

The Cambrian and Silurian rocks found in Wales were named for Cambria (the former name of Wales, from the Welsh *Cymru,* pronounced "koomri") and the Silures, a warlike tribe that lived in South Wales in Roman times. In 1835 Adam Sedgwick (1785–1873) and Sir Roderick Impey Murchison (1792–1871) jointly published a paper entitled "On the Silurian and Cambrian Systems, Exhibiting the Order in which the Older Sedimentary Strata Succeed Each Other in

THE STRATIGRAPHIC COLUMN IN THE LATE NINETEENTH CENTURY		
Tertiary	Cainozoic	Recent
		Pleistocene
		Pliocene
		Miocene
		Eocene
		Paleocene
Secondary	Mesozoic	Cretaceous
		Jurassic
		Triassic
Transition	Paleozoic	Permian
		Carboniferous
		Devonian
		Silurian
		Ordovician
		Cambrian
Primary		Precambrian

England and Wales." When they traced the boundary between Murchison's Silurian and Sedgwick's Cambrian farther afield, however, they were found to overlap. This led to a bitter dispute between the two friends that was not resolved until 1879, when Charles Lapworth (1842–1920) proposed that the disputed area be allotted a new name, the *Ordovician,* after the Welsh tribe of the Ordovices. Murchison and Sedgwick also named the *Devonian* for rocks in the county of Devon, England.

By this time geologists had found it necessary to introduce a "Transition" series of strata between the Primary and Secondary series. The Primary series then comprised all of the Precambrian, and the Transition series began with the Cambrian and ended with the Permian. Murchison named the *Permian* in 1841 following a tour of Russia. The name referred to the ancient kingdom and modern city of Perm in the Ural Mountains.

The history of these two very famous geologists illustrates the state of scientific education in Britain at the start of the 19th century. Sedgwick had studied mathematics at Cambridge University, was ordained a priest in 1818, and in the same year was appointed Woodwardian Professor of Geology, a post he held for the rest of his life. At the time of the appointment he knew nothing of geology. Murchison was a soldier whose wife persuaded him to pursue his long-standing interest in geology as an alternative to fox hunting. He became a full-time geologist in 1855, when he was appointed director of the Geological Survey.

By the 1840s the outline of the stratigraphic column was more or less complete. By the end of the century it had a recognizably modern form.

USING FOSSILS TO DATE ROCKS

By the late 18th century geologists understood that sediments deposited on the ocean floor accumulated as horizontal strata and that pressure might later compress them into rocks. Sedimentary rocks formed in distinct strata, one on top of another with the oldest at the bottom and the youngest at the top. Recognition of this allowed them to represent the strata as a stratigraphic column.

There was a problem, however: Not every part of the world had lain beneath the ocean, and some regions that had had long since lost their sedimentary rocks to erosion. The Earth's crust is dynamic, and its upheavals are capable of turning entire stacks of strata upside down. Similar sediments accumulate in different places at different times, so the discovery of a particular type of sedimentary rock in one place does not imply that an apparently identical rock found a long distance away formed at the same time. Werner's idea that each sedimentary rock possesses its own suite of minerals had been found to be untrue, removing what Werner had believed to be a means of determining the relative ages of the strata. This left geologists with the knowledge that similar rocks in different places may well be of vastly different ages. So how could the rocks of the stratigraphic column be arranged in a reliable chronological sequence?

Fossils provided the links that gave stratigraphy a historical dimension. Anatomists and zoologists were applying their knowledge of modern animal physiology to the reconstruction of animals they knew only from fossils. This led in one direction toward methods of classification. In *Handbuch der Naturgeschichte* (Handbook of natural history), published in 1779, Johann Friedrich Blumenbach (1752–1840), professor of medicine at the University of Göttingen, proposed that animals be classified according to bodily structures associated with the animals' functions; for example, he pointed out that humans have an anatomy that is clearly designed for walking upright (bipedalism). This distinguishes them from other primates, which are not bipedal. Blumenbach went further. Certain parasites are found in domestic pigs but not in wild pigs or in any other animals. Blumenbach concluded from this that the parasites could not possibly have existed prior to the domestication of the pig and, therefore, that they did not exist when the world was created. He studied fossils and found that some were so different from modern animals that he was unable to classify them. All of this suggested that plants

CUVIER AND BRONGNIART: THE SCIENTISTS WHO STUDIED THE FOSSILS OF THE PARIS BASIN

Georges-Léopold-Chrétien-Fréderic Dagobert was born on August 23, 1769, in Montbéliard in Württemberg, which was then a German principality. The family was descended from Huguenots, Protestants who had been forced to leave France during the reign of Louis XIV (1643–1715) and had settled in Switzerland. Dagobert's father was a Swiss national, but at the time of Georges's birth he was serving as a soldier in the French army.

After studying natural history at the University of Stuttgart, Dagobert spent six years as tutor to an aristocratic Protestant family in Normandy. In 1795 he moved to Paris to take up a position as assistant to the professor of comparative anatomy at the Museum of Natural History. In 1799 he was appointed professor of natural history at the Collège de France, and in

1802 he became a professor at the Jardin des Plantes (Botanical Garden). The following year he was appointed permanent secretary of physical and natural sciences at the Institut National. Napoléon made him a councillor of state and in 1808 commissioned him to examine the state of education in France, and when the Catholic Bourbons returned to the throne following the defeat of Napoléon, Dagobert served in the cabinet of Louis XVIII. In 1831 King Louis-Philippe ennobled him, and he became Baron Cuvier.

Cuvier used his deep understanding of comparative anatomy and its relationship to the way animals live to classify fossil vertebrate species. The story is told of how a student dressed in a devil's costume broke into Cuvier's room in the middle of the night, crying "Cuvier, Cuvier, I have come to eat you." Cuvier glanced at his assailant and said: "All and animals had become extinct at certain times in the past, and other species had replaced them. Blumenbach believed that species were created, but that creation was not a single, unique event. Species could disappear, and from time to time new species were created. Blumenbach was one of the earliest thinkers to recognize that the Earth, and all its plants and animals, had a very long history.

The French zoologist Georges Cuvier, a professor of comparative anatomy at the Museum of Natural History in Paris, is usually credited with having founded modern paleontology, the study of fossils. Cuvier was highly skilled at interpreting fossils and using fossilized bones to reconstruct the appearance of extinct animals. He collaborated with the mineralogist and naturalist Alexandre Brongniart (1770–1847) on a study of the sedimentary rocks of the Paris Basin (see sidebar). These were deposited during the Tertiary subera, which geologists now know lasted from 65.5 million years

creatures with horns and hooves are herbivores. You can't eat me." Then he went back to sleep. Cuvier also divided animals into four phyla (Vertebrata, Mollusca, Articulata, and Radiata).

Cuvier believed that the history of Earth involved a series of catastrophes (revolutions) in which whole populations of living organisms had been destroyed and their places taken by migration or the creation of new species. This catastrophist view of history explicitly rejected the concept of evolution. Cuvier died in Paris on May 13, 1832.

Alexandre Brongniart was born in Paris on February 5, 1770, the son of the architect Alexandre-Théodore Brongniart (1739–1813). He studied chemistry under Antoine-Laurent Lavoisier (1743–94); mineralogy at the École des Mines, where he also taught; and medicine at the École de Médicine. He then joined the army as a pharmacist but was imprisoned for a minor political offense. After

his release, in 1797 Brongniart was appointed professor of natural history at the École Centrale des Quatre-Nations in Paris. In 1800 Brongniart was made director of the Sèvres Porcelain Factory, a post he held until his death. He worked hard to improve the art of enameling and made Sèvres Europe's leading enamel factory. He was elected a member of the Academy of Sciences in 1807. From 1822 until his death he was professor of mineralogy at the Museum of Natural History.

In a paper published in 1800, "Essai d'une classification naturelle des reptiles" (Essay on a natural classification of reptiles), Brongniart classified reptiles into four orders: Batrachia (amphibians), Chelonia (turtles), Ophidia (snakes), and Sauria (lizards). His Batrachia are now a separate class, Amphibia. He also made the first classification of trilobites. Brongniart and Cuvier began their study of the Paris Basin strata in 1804. Brongniart died in Paris on October 7, 1847.

ago to 1.81 million years ago, so in geological terms the Paris rocks are fairly recent. They are also rich in fossils. As they worked, the two became convinced that distinctively different groups of organisms had lived during the time each stratum of sediment accumulated and, consequently, that the fossils they contained identified the relative ages of the strata. Cuvier identified the fossils, and Brongniart used the resulting information to arrange the geologic formations of the Tertiary into the Paleogene and Neogene periods (see "Geologic Timescale," pages 151–152).

The diagram shows why this is important. Upheavals in the crustal rocks can tilt strata so they are no longer horizontal and may even be tilted until they are vertical. They can turn columns of strata

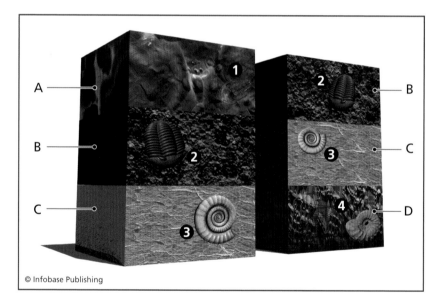

© Infobase Publishing

Fossils and rock strata. There are three strata in each column. The rocks are similar in each stratum, but the fossils they contain make it possible to determine their relative dates. The uppermost stratum in the column on the right contains fossils of type ii, found in stratum B in the left column. Therefore, the uppermost stratum in the righthand column (B) must be identical to the middle stratum (B) in the lefthand column. Similarly, stratum C, containing fossils of type iii, occurs at the base of the lefthand column and as the middle stratum in the righthand column. Stratum A, with fossil type i, is absent from the righthand column, and stratum D, with fossil type iv, is absent from the lefthand column.

upside down and strip away strata. But if a stratum contains fossils of a particular organism and fossils of that type are absent from other strata in the column, then all rock strata containing those fossils must be of similar age, regardless of where the rocks are found or how they are oriented. Some strata may contain no fossils, but their ages can be determined by the ages of the strata above and below them. The drawing shows two columns, presumably found in widely separate locations, each with three strata. Fossils of the type (i) found in the uppermost stratum (A) of the drawing on the left are absent from all the strata in the column on the right, but fossils of type ii, found in the middle stratum (B) on the left are present in the topmost stratum on the right. Fossils of type iii occur in stratum C, at the base of the left column and in the middle of the right column, and fossils of type iv occur in stratum D, which is present in the right column but absent from the left. It follows that strata B and C in one column are of the same age as B and C in the other column, regardless of their position in either column.

Applied on a much wider scale this technique allowed geologists to arrange sedimentary rock strata in a coherent chronological sequence. In 1811 Cuvier and Brongniart published their findings in a paper, "Essai sur la géographie minéralogique des environs de Paris, avec une carte géognostique et des coupes de terrain" (Essay on the mineralogical geography of the Paris area, with a geognosic [geologic] map and cross-sections of the terrane) in the *Annales d'Histoire Naturelle,* stimulating other geologists to embark on field trips to make surveys of their own areas.

WILLIAM SMITH—AND HIS GEOLOGIC MAP

Cuvier and Brongniart included a geologic map in their paper on the fossil-bearing strata of the Paris Basin. European nations in the 18th century were increasingly aware of the importance of accurate, reliable maps to navigate around and control their territories both at home and overseas. They also recognized the importance of the sciences of botany, zoology, and geology, and of the fact that scientists working in those disciplines needed specialized maps showing the distributions of plants, animals, and rock formations.

César-François Cassini de Thury (1714–84) supervised the production of the first comprehensive topographical map of France.

Work on the map began in 1744 and continued until Cassini's death in 1784. In 1766 the government commissioned Jean-Étienne Guettard (1715–86) to produce the first geological map of the country based on those Cassini topographical sheets that had so far been published (see sidebar).

It was a time of change and rapid advance. In the decade following the French Revolution of 1789, the government renamed the Royal Garden as the Museum of Natural History and made it a teaching

JEAN-ÉTIENNE GUETTARD—AND HIS MAPS

In the course of the 18th century the French government took advantage of the advanced surveying and cartographical techniques that were becoming available to commission detailed maps of the entire country. These were made under the supervision of the French astronomer and geodesist (geologist) César-François Cassini de Thury (1714–84), the grandson of the astronomer G. D. Cassini (see "Oblate or Prolate?" pages 14–16) and the son of Jacques Cassini (1677–1756), who was also an astronomer. The Cassini maps were accurate and of very high quality.

In 1746 the naturalist and mineralogist Jean-Étienne Guettard (1715–86) presented the first-ever mineralogical map of France to the Académie des Sciences, and in 1766 Henri-Léonard-Jean-Baptiste Bertin (1720–92), the French minister responsible for mines, commissioned Guettard to produce the *Atlas et description minéralogiques de la France,* based on Cassini's topographical maps. This was the most ambitious mapping project of its day. Guettard's assistant would be the chemist Antoine-Laurent Lavoisier (1743–94). Later the supervision of the work was passed to another mineralogist, Antoine-Grimoald Monnet (1734–1817). Disputes among surveyors and finan-

cial crises plagued the project from the start, and many maps were never completed. Those that were finished appeared in 1780. Contour lines to link points of similar elevation had not yet been invented, so maps in the *Atlas* indicated elevation by shading. Despite this limitation, the maps were an impressive achievement, showing considerable detail.

Guettard was born on September 22, 1715, at Étampes, in the Île de France, south of Paris. He studied botany under Bernard de Jussieu (1699–1777) and mineralogy under the physicist René-Antoine Ferchault de Réaumur (1683–1757), and for a time in 1741 he was curator of Réaumur's natural history collection. Guettard qualified as a doctor of medicine and in 1742 became a member of the Faculty of Medicine of Paris. From 1747 until 1752 he was doctor-botanist to Prince Louis, duc d'Orléans, and became keeper of his natural history collection. Guettard was the first scientist to study the exposed bedrock of the Paris Basin, and he identified several fossils found in that area. He was also first to recognize that many rocks in the Auvergne region were volcanic in origin and he identified several of the Auvergne hills as extinct volcanoes. He died in Paris on January 6, 1786.

institution as well as a museum. The École des Mines taught courses in geology as well as mining engineering, and in Germany there also were mining schools offering similar courses. Alexander von Humboldt studied at the most famous of these, in Freiberg. Geological societies were founded to facilitate the exchange of information among geologists.

In Britain the geologist Henry Thomas de la Beche (1796–1855) began to compile a geologic map of Devon, in southwestern England, financed by the government. He had written extensively on the stratigraphy of Devon and Dorset and had conducted field surveys in southwestern Wales and in Jamaica. Born simply as Henry Beche, he had been educated at a military academy and served as an officer in the Napoleonic Wars, but when he left the army, he added the *de la* to his name and became an amateur geologist, joining the Geological Society of London in 1817. In the 1830s he conceived a plan to produce a geological map of the whole of Britain and used his friendship with the prime minister, Sir Robert Peel (1750–1830), to establish the Geological Survey, with himself as its first director.

De la Beche was not involved in the production of the first British geological map, however. This appeared in 1815 with the publication of *A Delineation of the Strata of England and Wales,* followed between 1816 and 1824 by *Strata Identified by Organized Fossils* and *Stratigraphical System of Organized Fossils,* which was a catalog. There were also various charts and complete maps of 21 counties. These were all the work of William Smith (1769–1839), affectionately nicknamed "Strata Smith." Smith was born on March 23, 1769, into a farming family in Churchill, Oxfordshire. His father died when he was seven years old, and William was raised by an uncle. He received only a rudimentary education.

In 1787 Smith went to work for Edward Webb, a land surveyor at Stow-on-the-Wold, in the neighboring county of Gloucestershire, and he spent the next eight years learning the skills of that profession, first with Webb and later in Somerset with the Somerset Coal Canal Company. He surveyed the rock strata on an estate located above part of the Somerset coalfield, where he realized that the order of the strata was the same wherever the rocks occurred and that he could identify each stratum by the fossils it contained. This meant the order of strata was predictable over a large area. His conclusion that each stratum can be identified by the particular assembly of fossils it contains and that the fossils succeed one another in the strata in a definite and regular order is now known as the *law of faunal succession.*

At that time a network of canals was being built across England to link navigable rivers and provide a means for the cheap transport of coal and other bulky goods. From his work surveying the coalfield, in 1794 Smith was appointed resident engineer to the Somerset Coal Canal Company, with the tasks of surveying the routes for two new canals and then supervising their construction. First the company sent him on a tour lasting almost two months that took him to Newcastle in the north, Shropshire, Wales, and finally to Bath, back in Somerset.

Having planned the most economic route for each canal, Smith moved on to the next phase. Teams of workers known as "navigators," commonly shortened to "navvies," worked with dynamite, pick,

A modern geologic map of the surface formations in an area of Maine uses colors to indicate rock formations and sedimentary deposits. *(Biospheric Sciences Branch NASA/GSFC)*

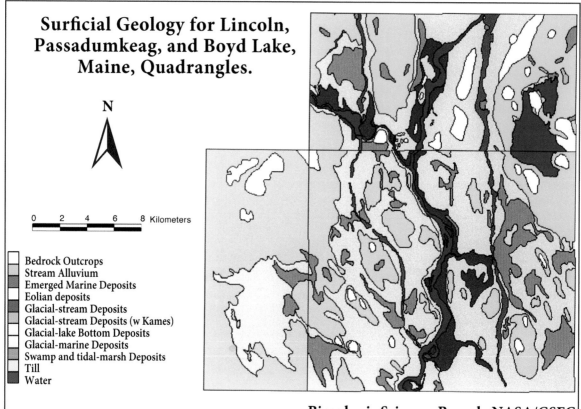

Surficial Geology for Lincoln, Passadumkeag, and Boyd Lake, Maine, Quadrangles.

N

0   2   4   6   8 Kilometers

Bedrock Outcrops
Stream Alluvium
Emerged Marine Deposits
Eolian deposits
Glacial-stream Deposits
Glacial-stream Deposits (w Kames)
Glacial-lake Bottom Deposits
Glacial-marine Deposits
Swamp and tidal-marsh Deposits
Till
Water

Biospheric Sciences Branch NASA/GSFC

and shovel to cut the canals. Until the canals were completed and filled with water, their work exposed the strata, and if those strata contained fossils, Smith, the engineer, was on hand when they were discovered. He was fortunate in that these two canals cut through gently inclined strata, some of which were rich in fossils. Smith wrote a first account of his work and discoveries in 1796.

He lost his job in 1799 when his work on the canals was completed and took a number of engineering jobs in different parts of England and Wales. This gave him opportunities to study exposed rocks wherever he could find them. It was his fascination with the strata that earned him his nickname.

Smith produced his first large-scale geologic map, of the area around Bath, Somerset, in 1799, using colors—which he applied by hand—to denote different rock formations. This is still the way geologic maps are produced, as the illustration shows, with different maps to show different aspects of the structure and deposits of a region.

In 1801 Smith produced the first rough sketch of his 1815 geologic map of England. In addition to rock formations, the 1815 map used symbols to indicate the location of roads, tramways, canals, tunnels, coal mines, saltworks, some chemical factories, and lead, copper, and tin mines. In 1817 Smith published a geological section showing a cross-section through the rocks all the way from Snowdon in North Wales to London, a distance of about 194 miles (312 km).

Smith financed his travels himself, and his maps became well known, but it was not long before they were being copied and sold more cheaply than the prices he needed to charge in order to cover his costs. At the same time academic scientists often treated him with disdain because of his humble origins and lack of formal education. He felt snubbed, and the pirating of his work eventually bankrupted him. He was imprisoned for debt and on his release on August 31, 1819, arrived home only to find a bailiff waiting to seize his home and all his property.

Smith's contribution to geology was finally recognized in January 1831 when the Geological Society of London awarded him the Wollaston Medal, given to "geologists who have had a significant influence by means of a substantial body of excellent research in either or both pure and applied aspects of the science." The medal was funded by a bequest from William Hyde Wollaston (1766–1828), and Smith was its first recipient. During his address at the award ceremony, Adam Sedgwick (see "Layers of Rock," pages 126–129) described Smith as "the father of English geology." The government awarded Smith a pension of £100 (about $5,000 in today's money) a year for the rest of his life. In 1835 he received a doctorate of law during the meeting of the British Association for the Advancement of Science, held that year in Dublin, and in 1838 he was appointed to the committee charged with choosing stone for building the new Houses of Parliament. He died in Northampton on August 28, 1839.

CATASTROPHISM—AND WILLIAM BUCKLAND

William Buckland (1784–1856) and his friend William Daniel Conybeare (1787–1857) were sons of Church of England clergymen who had been sent to Oxford to gain a degree as the necessary prerequisite for ordination. Then, as was customary, they stayed on after graduating to wait either for a fellowship to their Oxford colleges or for a position as a curate (junior parish clergyman) to become vacant. They were destined to become clergymen but were keenly interested in geology. Both attended a course on mineralogy, and they devoted their free time to geological field excursions around Oxford and during their vacations studied rock formations much farther away.

Buckland graduated in 1804, obtained his master's degree in 1808, and was ordained quickly to allow him to accept a fellowship to Corpus Christi College. In 1813 he became reader in mineralogy at Oxford. He was elected a fellow of the Royal Society in 1818, and in 1819 he was appointed Oxford's first-ever reader in geology. In his inaugural lecture, delivered on May 15, 1819, and published the following year, Buckland sought to reconcile the geological evidence for Earth's history with the biblical account. Buckland became president of the Geological Society in 1824, and in 1825 he resigned his Corpus Christi fellowship to become a clergyman at Stoke Charity, a village in Hampshire. In 1845 he became dean of Westminster. Conybeare left Oxford in 1814 to become a country clergyman and to marry, but he continued to pursue his geological interests in his free time and remained in close touch with Buckland.

Early in his career, Buckland toured Europe collecting specimens for the Ashmolean Museum at Oxford. This brought him into contact with many prominent scientists, including Cuvier (see "Using Fossils to Date Rocks," pages 129–133). Cuvier's study of fossils had led him to conclude that in the course of its history the Earth had experienced many violent "revolutions" that had wiped out most of the existing animals, which had been replaced by new forms. His "revolutions" were catastrophic events, and Cuvier's interpretation of the fossil record came to be called catastrophism. Buckland was also a keen and skilled paleontologist. He discovered the fossil bones of a large reptile, which he named *Megalosaurus* meaning "great lizard," and in 1826 he discovered the oldest human remains known in Britain up to that time. Buckland was strongly influenced by Cuvier but adapted Cuvier's catastrophic theory to introduce a religious component.

Cuvier's theory was not religious. He believed many catastrophes might have involved widespread flooding, but he never referred any of them to the biblical flood, and he never suggested that the world had been repopulated by divine creation following mass extinctions. He did believe that the catastrophes were separated by very long periods when the Earth was stable. This implied that the Earth was several million years old.

Buckland had no difficulty with this, but to reconcile the evidence with the Bible he needed to establish that the most recent catastrophe had been a worldwide deluge—Noah's flood—and he found what he thought was the evidence he needed in caves. The Napoleonic Wars finally ended in 1815, and in 1816 Conybeare, Buckland, and their friend George Bellas Greenough (1778–1855) embarked on a tour of Europe, partly financed by Greenough, who was independently wealthy. They met other geologists, examined collections of geological specimens, and while they were in Bavaria, they visited caves in which earlier workers had found fossil bones. One such cave was near the village of Gailenreuth. Cuvier had identified the bones found in the Gailenreuth cave as those of giant bears, belonging to a species now extinct. Some scientists, including Cuvier, believed the bears had used the cave as a den; others thought the bears had died outside and their bones had been swept into the cave by floodwater. No one knows what Buckland thought of the matter at the time, but years later he suggested that the bears had been living in the cave when a sudden flood overwhelmed and drowned them.

Back in England Buckland investigated a cave at Kirkdale, near Pickering in Yorkshire. Mud filled the cave, but embedded in the mud were fossil bones bearing teeth marks. Buckland found that these matched precisely the marks on bones he had obtained from the hyena cage at London Zoo. The cave had been a hyena den, and Buckland concluded that the period when the hyenas lived there had ended with a flood that killed its occupants and filled the cave with mud. He assumed that this flood had been worldwide, and in 1823 he published his findings under the title *Reliquiae Diluvianae, or, Observations on the Organic Remains attesting the Action of a Universal Deluge.* His identification of the teeth marks was accurate, and a flood was the most reasonable explanation for the mud filling the cave, but he had no evidence that the flood was universal. Nevertheless, he was satisfied he had reconciled geology and religion, although the two were in agreement only over that particular flood. Buckland held that this universal flood was but the latest in a long series of catastrophes extending over many millions of years.

A widespread ice age is a different type of catastrophe, and Buckland became interested in the work of Agassiz (see "Ice Ages," pages 122–126). In 1838 he went to Switzerland to meet Agassiz and see the evidence for himself, returning convinced that Agassiz was correct. He then reinterpreted observations he had made in Britain and attributed to the biblical flood. He now realized they were caused by ice.

Buckland combined his geological work with his duties as a minister at Westminster and also in the town of Islip, near Oxford. In 1848 the Geological Society awarded him the Wollaston Medal. Late in 1849 he fell ill and grew increasingly incapacitated until his death, on August 24, 1856.

Catastrophism fell out of fashion (see "James Hutton, Plutonism, and Uniformitarianism," pages 144–149), and the idea of repeated floods gave way to the idea proposed by the French geologist Léonce Élie de Beaumont (1798–1874) of periodic convulsions of the rocks as the Earth slowly cooled (see "Cooling and Crumpling," pages 155–157). For much of the 20th century geologists believed that the Earth's history unfolded slowly through processes that operated constantly. That view was challenged in 1979 by the publication of evidence that about 65 million years ago the Earth had been struck by an asteroid, resulting in a worldwide catastrophe (see sidebar). Nowadays catastrophism is once more acceptable in its new, modified form.

MODERN CATASTROPHISM—AND THE DEATH OF THE DINOSAURS

Far from Earth, between Mars and Jupiter there is a ring composed of countless millions of rocks of varying sizes. This is the asteroid belt, and deep inside it there is a group of rocks known as the Baptistina asteroid family. At one time these rocks were joined together, forming a single asteroid approximately 105 miles (170 km) in diameter, but about 160 million years ago that body was shattered in a collision. Influenced by the gravitational attraction of neighboring bodies, little by little the Baptistina fragments left the asteroid belt, some on orbits that brought them close to Earth. About 109 million years ago a fragment about 2.5 miles (4 km) across struck the Moon, forming the 53-mile- (85-km-) diameter crater Tycho. About 65 million years ago a fragment about six miles (10 km) across struck Earth.

In 1979 Walter Alvarez, Luis W. Alvarez, Frank Asaro, and Helen V. Michel published in volume 11 of *Geological Society of America: Abstracts with Program* an account of their discovery of traces of two metals, iridium and osmium, in a thin layer of clay at Gubbio, Italy, and the implications of that discovery. The following year they told the story again in *Science* in a paper entitled "Extraterrestrial Causes for the Cretaceous—Tertiary Extinction." Iridium and osmium are very rare in the rocks of the Earth's crust but much more common in extraterrestrial dust and other material that has survived from early in the formation of the solar system. The Alvarez team found that the Gubbio clay layer contained about 30 times more of the two metals than the material above and below that level, and the clay was dated at about 65 million years old. Later, similar enrichment was discovered in clays of the same age at Stevns Klint, Denmark, and Woodside Creek, about 25 miles (40 km) from Wellington, New Zealand. The enrichment was global in extent, and the most plausible explanation for its presence was that it had arrived from space. Relating calculations of the total amount of iridium and osmium with the average concentration in extraterrestrial material led the scientists to conclude that the enrichment resulted from the impact and destruction of an extraterrestrial body about six miles (10 km) in diameter that struck the Earth 65 million years ago traveling at about 12 miles per second (20 km/s). An impact of that magnitude would leave a large crater, but several years passed before the crater was discovered, beneath Chicxulub, on the Yucatán Peninsula, in Mexico.

The boundary between the end of the Cretaceous and beginning of the Paleogene periods, dated at 65.5 million years ago, is marked by the disappearance of fossils of a wide range of animals that are present before the boundary but absent after it. A mass extinction of animals occurred at that time, but until the Alvarez discovery its cause had been a mystery. Today most geologists and paleontologists accept that the impact of a body from outer space, releasing an amount of energy equivalent to the detonation of about 100 trillion (10^{12}) tons of TNT, was responsible. The letters *K* (from the German

(continues)

(continued)
name for the Cretaceous, *Kalk,* meaning "chalk") and *T* for the start of the Tertiary subera are commonly used to identify the impact and extinction as the K/T boundary event.

About 215 million years ago the first dinosaurs appeared on Earth. The group flourished throughout the Jurassic and Cretaceous periods. They were at their peak in the later Jurassic, which is when the Baptistina asteroid was broken into fragments, one of which would collide with the Earth 95 million years later. Already their fate was sealed, for the dinosaurs were among the animals that perished in the K/T boundary event.

NEPTUNISM

The idea that on one or more occasions the whole Earth has been flooded is called *Diluvialism.* This is not the same as Neptunism, the theory that the rocks of the Earth's crust formed under water. Neptunism is often associated with Werner (see "Abraham Gottlob Werner—and the Classification of Rocks," pages 114–117), but the theory did not originate with him, and he never made any such claim. Indeed, he was not concerned with geological theory so much as with practical descriptions and classifications of rocks and minerals. His Neptunism merely accepted a widely held view of the time.

Diluvialism and Neptunism were not always so widely separated, however. In 1681 Thomas Burnet (1635–1715), later to become Clerk of the Closet at the court of King William III of England, published *Telluris Theoria Sacra* (*Sacred Theory of the Earth*), which became very popular. The English translation appeared between 1684 and 1689.

Burnet maintained that beneath the Earth's surface there are vast chambers filled with water equivalent to the volume of nine oceans. When these erupted in the distant past, they caused the biblical flood. He suggested in the first edition of his book that the Earth's crust is the firmament mentioned in Genesis. The suggestion was removed from later editions, but Hooke (see "Robert Hooke, Who Showed That Long Ago Britain Lay Beneath the Sea," pages 99–102) shared this view. Having accounted for the volume of water that would be needed to produce a universal flood, Burnet went on to speculate about how the Earth acquired what he believed to be its form. It began with chaos, he said, out of which materials settled according to their density, the densest first. The heaviest materials formed the core of the Earth and were enclosed in water. Lighter materials and air then settled on top, forming a completely smooth surface. In those days the entire world enjoyed a climate similar to a perpetual English spring, with no seasons. Then the Flood occurred, due to entirely natural causes, and the postdiluvial world was left ruined. In 1692 Burnet published another book in which he suggested that the six days during which Genesis states the Earth was created should be interpreted allegorically rather than literally. This caused such strong opposition that the king was forced to remove him from his position at court.

John Woodward (1665–1728) aimed to improve on Burnet's explanation for the origin of the Earth's crust. Woodward agreed that a vast ocean had once covered the entire Earth. He maintained that the water of the ocean contained a complicated mixture of substances, and as the sea level gradually fell, these substances were precipitated in order of their specific gravities, the heaviest first. These substances formed the sedimentary rocks that Lehmann would later term *Secondary* (see "Abraham Gottlob Werner—and the Classification of Rocks," pages 114–117). Woodward believed that the world ocean resulted from the Flood described in Genesis and that the rocks formed in the timescale indicated in Genesis. Woodward was very influential, and many educated people agreed with his account of the formation of rocks.

Woodward was born on May 1, 1665, in Derbyshire, England, and when he was 16, he went to London to study medicine under the physician to Charles II, Peter Barwick. He was appointed professor of physic (medicine) at Gresham College, London, in 1692, was elected a fellow of the Royal Society in 1693, and in 1702 he became a fellow of the Royal College of Physicians.

Although he was employed as a physician, Woodward had been interested in botany and natural history since his student days and was especially attracted to fossils. Through close examination of fossil-bearing rocks he came to realize that the rocks formed distinct strata and that the fossils were of marine organisms, indicating that the rocks had originated on the seabed. Woodward died on April 25, 1728, and was buried in Westminster Abbey.

Woodward's theory of rock formation was widely accepted, but it was modified and amended during the course of the 18th century. It was soon noted, for example, that the ocean's ingredients could not have been precipitated in order of their specific gravity because dense rocks were often found lying above lighter rocks. Scientists also noted the rate at which fine sediments accumulate, and when they compared that to the thickness of some of the sedimentary strata, it was obvious that the rocks could not possibly have formed in the short time allowed in Genesis and which Woodward accepted. Nevertheless, the central idea survived, that the secondary rocks formed by sedimentary processes on the floor of an ocean that covered the entire world. In time this was extended to include the primary rocks, such as granite.

It was a powerful theory, and there seemed to be much evidence to support it. If rock particles accumulated as the global sea level fell, then the oldest rocks should be found at the highest elevations—on mountains—and younger rocks at lower levels. The illustration explains why this is so. Studies in the Alps, where granite occurs near the peaks and sedimentary strata lower down, tended to confirm this. As the rocks formed, the removal of their ingredients would alter the composition of the ocean; consequently, rocks of particular types should occur only at certain levels in the stratigraphic column. Studies seemed to confirm that, too.

This was the basic structure of the Neptunist theory that Werner revived at Freiberg. Its principal British exponent was probably Robert Jameson (1774–1854), Regius Professor of Natural History at the University of Edinburgh from 1804 until his death, who went to the Freiberg Mining Academy in 1800 and spent a year there studying under Werner. Jameson founded and until 1850 was president of the Wernerian Natural History Society. In 1850 his health was beginning to fail and interest in the society was waning. He died in Edinburgh on April 19, 1854. As a very young man (he was only 16), Charles Darwin (1809–82) studied under Jameson at Edinburgh.

Eventually, though, even more compelling evidence gave support to a rival theory: Plutonism and the theory of Uniformitarianism that Plutonism implied. Toward the end of his life Jameson renounced Wernerian Neptunism in favor of Plutonism.

JAMES HUTTON, PLUTONISM, AND UNIFORMITARIANISM

Plutonists argued that rocks form when molten magma or lava cools and solidifies. If the new rock is exposed above ground, erosion commences immediately and in time the rock is worn away.

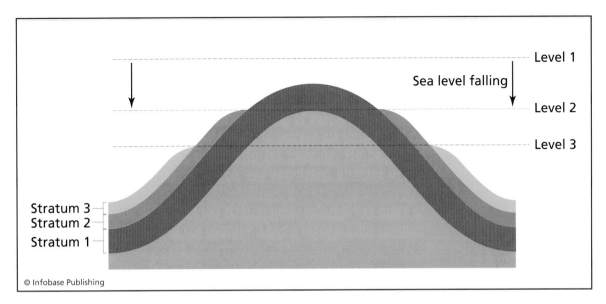

Level 1

Sea level falling

Level 2

Level 3

Stratum 3
Stratum 2
Stratum 1

© Infobase Publishing

Neptunism. In the Neptunist theory rocks formed by precipitation from an ocean that once covered the entire Earth, during a time when the global sea level was falling. The oldest rock strata should therefore occur at the highest elevations, and the rocks should be progressively younger at lower levels. As sea level falls from level 1 to level 2, the rocks form stratum 1; these are the oldest, because they formed first. At level 2 the rocks form the younger stratum 2, and at level 3 they form the youngest stratum 3.

The particles produced by erosion form sediments, which become sedimentary rocks. These are elevated, tilted, and eroded in their turn. Meanwhile, further volcanism produces a constant supply of fresh rock so that volcanism followed by erosion makes up an endless cycle.

The Plutonic theory was first proposed in 1785 by James Hutton (1726–97), a Scottish natural philosopher, in a paper titled "The System of the Habitable Earth with Regard to its Duration and Stability," that his friend the chemist Joseph Black (1728–99), delivered on his behalf to the recently formed Royal Society of Edinburgh. Illness prevented Hutton from presenting the paper in person. A fuller version, with the title *Theory of the Earth or an Investigation of the Laws Observable in the Composition, Dissolution, and Restoration of Land upon the Globe,* was published in 1788 in the *Transactions* of the society. *Theory of the Earth* appeared in 1795 as a two-volume work. A third volume, containing most of the evidence Hutton had assembled, was lost and not published until 1899.

Hutton's theory aroused little interest in his own lifetime and became widely accepted only after Charles Lyell (1797–1875) had promoted it 50 years later (see sidebar). It then became the generally accepted description for the history of the rocks at or near the Earth's surface, and Hutton is often called the founder or father of modern geology. There were two reasons for the theory's initially cool reception. The first was that Hutton did not write clearly, and his book was difficult to read and understand. It was not until 1802 that the Huttonian theory appeared in a more accessible form, in *Illustrations of the Huttonian Theory of the Earth*, written by Hutton's friend John Playfair (1748–1819), then professor of mathematics and natural philosophy at the University of Edinburgh.

The second reason was more substantial: The theory was not scientific. Hutton described processes that were intended to ensure the continued habitability of the Earth for humans, making the Earth a perpetual-motion machine built and operated by God with the purpose of maintaining the habitability of the Earth. The cycles of formation and erosion led him, in his original paper, to make his most famous statement: "The result, therefore, of this physical inquiry is, that we find no vestige of a beginning, no prospect of an end." But Hutton had formulated the theory first and sought evidence to support it later, and that is not science. Once he began looking for evidence, however, he soon found it. In 1785 he visited Glen Tilt, in the Cairngorm Mountains of the Scottish Highlands, where he discovered granite intruding into metamorphic schists. This showed that the granite had once been molten and that it had not crystallized from solution. He found a similar intrusion in the Salisbury Crags, near the center of Edinburgh, and in several other locations.

Other evidence came from overseas. Sir William Hamilton (1730–1803), the British ambassador to Naples and a distinguished antiquarian and volcanologist, was convinced that the volcanoes around the Mediterranean had been active since long before the beginning of human history. In 1769 he visited Mount Etna, one of the world's most active volcanoes, as the illustration on page 149 shows, but quiet during his examination of it. He noted that the flanks of the volcano were dotted with smaller cones produced by eruptions below the summit. Some of these had occurred in historic times, but others were covered with trees, and he concluded that they resulted from

CHARLES LYELL—AND MOUNT ETNA

In 1828 Charles Lyell visited Mount Etna in Sicily, the largest volcano in Europe. As a student Lyell had attended lectures by William Buckland (1784–1856) and was a convinced catastrophist, but he was unhappy with Buckland's attempts to link geology to events described in Genesis and in particular with the idea that the biblical flood was a genuine historical event. Buckland's theories also appealed to socially conservative Oxford academics whose views Lyell found uncongenial. Open to alternative ideas, Lyell was impressed by the writing of George Julius Poulett Scrope (1797–1876), who had published two books on volcanoes. Scrope showed that the extinct volcanoes in France had been formed not by single, large eruptions, but by a succession of eruptions separated by long periods of erosion. When Lyell examined Etna, he found that the mountain had grown slowly and only the most recent eruptions had occurred during recorded history. Later Lyell read the work of James Hutton (1726–97), realized he and Hutton had reached a similar conclusion, and revived and popularized the ideas of Hutton. Lyell is the person mainly responsible for making Plutonism and uniformitarianism the orthodox view of Earth history.

Lyell was born on November 14, 1797, in Kinnordy, Angus (then in Forfarshire), Scotland, but the family moved to Hampshire in the south of England while he was still a child. His father was a lawyer and amateur botanist, and Charles became interested in natural history, especially in butterflies and moths, and in geology. He studied rock formations on family holidays in Britain during his childhood and later in continental Europe.

Charles Lyell wrote *Principles of Geology*, probably the most influential geology textbook of the 19th century, bringing the uniformitarianism of James Hutton to a wide scientific readership. *(Hulton Archive/Getty Images)*

Lyell studied classics at Oxford University, then law, and qualified as a lawyer in 1822, but he was more interested in geology, attending Buckland's lectures. He was appointed professor of geology at King's College, London, in 1831. Lyell was knighted in 1848 and made a baronet in 1864. He died in London on February 22, 1875, and is buried in Westminster Abbey.

Lyell divided the Tertiary era (as it was then defined) into the Eocene, Miocene, and Pliocene

(continues)

(continued)

epochs, and he suggested that some of the Earth's rocks might be as much as 240 million years old—far older than most geologists of the time believed. His fame rests mainly on his writing, however, and he was unusual in that writing was his principal source of income. The three volumes of his *The Principles of Geology* appeared between 1830 and 1833 and remained a leading textbook, with 12 revised editions, until the last edition appeared in 1875. He was a close friend of Charles Darwin (1809–82), and it was Lyell and Joseph Hooker (1817–1911) who in 1858 presented to the Linaean Society the two papers on evolution by natural selection written independently by Darwin and Alfred Russel Wallace (1823–1913), although some years passed before Lyell accepted Darwin's theory.

eruptions in remote prehistoric times. Lyell followed in Hamilton's footsteps in 1828 and made similar observations.

Plutonism emphasized the importance of volcanism in the formation of rocks, but Hutton's observations also suggested a timescale. The processes he described were extremely slow, and all of them were still active. This led him to conclude that the origin of rock formations and landscape features can be explained by reference to observable natural processes. This theory is called Uniformitarianism and is summed up as: The present is the key to the past.

Hutton was deeply religious, and his theory aimed to show how God manages the world for the benefit of humans; nevertheless, he aroused fierce opposition from some who thought he encouraged atheism. The trouble was that his account of Earth's history allowed no place for the biblical flood or for the story of creation recounted in Genesis.

James Hutton was born in Edinburgh on June 3, 1726, the son of a merchant. He studied law and then medicine at the Universities of Edinburgh, Paris, and Leiden, qualifying as a doctor in 1749, although he never practiced. After graduating, Hutton spent 20 years working as a gentleman farmer in Berwickshire, in southern Scotland, and it was during this time that he developed an interest in mineralogy and geology. He returned to Edinburgh in about 1768, where he joined the intellectually glittering society of the Scottish Enlightenment and became very active in the newly formed Royal Society of Edinburgh. Hutton died in Edinburgh on March 26, 1797.

HOW THEY BUILT THE GEOLOGIC TIMESCALE

By the middle of the 19th century geologists were agreed about the division and naming of the rocks that made up the stratigraphic column. They could identify similar strata in widely different places by means of the fossils contained in them or in the strata above and below them. But two important questions remained unanswered: How old is the Earth, and how long were each of the episodes into which geologists had divided its history? Although the rocks had been arranged in order of their age, there was no reference against which their absolute age in years could be measured. The problem was insoluble, the questions unanswerable, but scientists agreed there had to be an upper limit to the Earth's age, because they believed that the Earth had once been molten and that it was still

Mount Etna, on the island of Sicily, is Europe's largest volcano. Its height varies owing to its eruptions, but it is about 10,900 feet (3,324 m). One of the world's most active volcanoes, it is in almost continuous eruption. This eruption was photographed on October 30, 2002. *(Marco Di Lauro/Getty Images)*

cooling. Cooling is a process that must have had a beginning, and if there were a means of determining when the cooling process began, it would reveal the age of the Earth.

There were at least two ways to achieve this, or so it seemed. Physicists estimated the rate at which the planet was cooling, measured its present temperature, and used these data to calculate the planet's age. In 1862 Britain's most eminent physicist, William Thomson (later Lord Kelvin), used this method to calculate an age for the Earth of not less than 24 million years and not more than 400 million years. In 1899 the Irish scientist John Joly (1857–1933) calculated the rate at which soluble salts were draining from the land into the oceans, compared this with the salinity of seawater, and concluded that the oceans were between 80 million and 100 million years old.

Geologists had great difficulty accepting these estimates, for their understanding of the rate of the sedimentary processes that lead to the formation of sedimentary rocks suggested the Earth was much older than even 400 million years. The breakthrough that would allow accurate absolute dating occurred in 1896, the year in which the French physicist Henri Becquerel (1852–1908) discovered radioactivity. As physicists began to study the new form of energy, they quickly realized that radioactive elements give off heat over an extremely long period. That heat, from the decay of radioactive elements present in rocks and, therefore, in the Earth's mantle, would greatly slow the rate at which the Earth was cooling. Ernest Rutherford (1871–1937), the New Zealand physicist, was one of the first to study this aspect of radioactive decay and one of the first to recognize its implication that the Earth is far older than 400 million years. Robert John Strutt, Lord Rayleigh (1875–1947) developed Rutherford's idea about the Earth's age, and in 1913 Strutt's student Arthur Holmes (1890–1965) began using the rate of radioactive decay to calculate the Earth's age, a technique known as *radiometric dating*.

Holmes used the decay of uranium, which eventually becomes lead. Physicists had discovered that radioactive elements emit radiation at a very regular rate. Although it is impossible to predict when an individual atomic nucleus (a *nuclide*) of a radioactive element will spontaneously transform into a different nuclide, with a release of radiation, the rate at which a large mass of nuclides does so can be measured with great precision. The transformation of a nuclide into a nuclide possessing different characteristics is known as decay,

THE GEOLOGIC TIMESCALE					
EON/ EONOTHEM	ERA/ERATHEM	SUB-ERA	PERIOD/ SYSTEM	EPOCH/ SERIES	BEGAN MA
Phanerozoic		Quaternary	Pleistogene	Holocene	0.11
				Pleistocene	1.81
	Cenozoic	Tertiary	Neogene	Pliocene	5.3
				Miocene	23.3
			Paleogene	Oligocene	33.9
				Eocene	55.8
				Paleocene	65.5
	Mesozoic		Cretaceous	Late	99.6
				Early	145.5
			Jurassic	Late	161.2
				Middle	175.6
				Early	199.6
			Triassic	Late	228
				Middle	245
				Early	251
	Paleozoic	Upper	Permian	Late	260.4
				Middle	270.6
				Early	299
			Carboniferous	Pennsylvanian	318.1
				Mississipian	359.2
			Devonian	Late	385.3
				Middle	397.5
				Early	416
		Lower	Silurian	Late	422.9
				Early	443.7
			Ordovician	Late	460.9
				Middle	471.8
				Early	488.3
			Cambrian	Late	501
				Early	542
Proterozoic	Neoproterozoic		Ediacaran		600
			Cryogenian		850

(continues)

THE GEOLOGIC TIMESCALE (continued)

EON/ EONOTHEM	ERA/ERATHEM	SUB-ERA	PERIOD/ SYSTEM	EPOCH/ SERIES	BEGAN MA
	Mesoproterozoic		Tonian		1,000
			Stenian		1,200
			Ectasian		1,400
			Calymmian		1,600
	Palaeoproterozoic		Statherian		1,800
			Orosirian		2,050
			Rhyacian		2,300
			Siderian		2,500
Archaean	Neoarchaean				2,800
	Mesoarchaean				3,200
	Palaeoarchaean				3,600
	Eoarchaean				3,800
Hadean	Swazian				3,900
	Basin Groups				4,000
	Cryptic				4,567.17

Source: International Union of Geological Sciences, 2004.

Note: Hadean is an informal name. The Hadean, Archaean, and Proterozoic Eons cover the time formerly known as the Precambrian. Tertiary had been abandoned as a formal name, and Quaternary is likely to be abandoned in the next few years, although both names are still widely used. Ma means "millions of years ago."

and the rate at which it decays is called the *decay constant.* From the decay constant it is possible to calculate the time it will take for exactly half of the nuclides of an element to decay. This is called the *half-life.* It is absolutely constant for every radioactive element, and every element has a different half-life.

Two isotopes of uranium, ^{238}U and ^{235}U are used in uranium-lead dating. ^{238}U has a half-life of 4,460 million years and decays to lead-206 (^{206}Pb). ^{235}U has a half-life of 704 million years and decays to lead-207 (^{207}Pb). All ^{206}Pb and ^{207}Pb result from the decay of uranium, so the age of a rock can be calculated from the proportions of the isotopes present in a rock. Uranium-lead dating is accurate for rocks up to about 4.6 billion years old, and there are now other techniques using radioactive decay to date rocks that do not contain uranium or lead. Radiometric dating methods have finally solved the problem of determining the age of the Earth. The planet is 4.657.17 billion years old.

Holmes described his findings in *The Age of the Earth,* published in 1913. The book included his geologic timescale, showing the duration of each division of the Earth's history. The International Commission on Stratigraphy (ICS), which is part of the International Union of Geological Sciences (IUGS), now produces the geologic timescale that all geologists use. The IUGS is the body that brings together the national geological societies of 118 nations. From time to time the stratigraphers of the ICS amend the geological timescale to take into account new discoveries. The table on pages 151–152 shows the current (2004) version.

How Do Mountains Rise?

In the late 17th century Thomas Burnet proposed in *Sacred Theory of the Earth* that the surface of the Earth had originally been perfectly smooth (see "Neptunism," pages 142–144). His idea proved popular and is very revealing. Burnet believed that the world's present landscapes of hills and valleys, high, rocky, snow-capped mountains, and dramatic coastal cliffs represented the ruin of the once-perfect surface.

Nowadays most people admire the grandeur of mountains. Their awesome beauty adorns calendars and travel brochures. They attract tourists and climbers. This is a modern attitude however. As recently as the 19th century mountains were seen as fearful places, wildernesses where travelers might easily become disoriented and lost, where a slip on wet rock might send someone plunging to his or her death over a precipice, where fierce beasts haunted the caves and shadows. Most important, at a time when the failure of a harvest often meant starvation and famines were common, mountains were places that produced no food.

Of course, geologists had to recognize the fact that mountains exist whether they liked them or not, and since they exist, they had to discover how they came to be. Certain types of volcanoes grow into mountains, but sediments are deposited in horizontal layers. Sedimentary rocks form as approximately level strata beneath the ocean floor. How can it be, then, that they are found near mountain peaks, complete with fossils of marine organisms?

This chapter explores the question of the origin of mountains and the route by which the answer was found. It begins with the realization that the interior of the Earth is hot and the supposed implications of that discovery.

COOLING AND CRUMPLING

If the Earth formed as a molten mass or became molten soon after it formed, the planet must have been cooling throughout its history. Its surface rocks are cool, but beneath them the interior of the Earth is still hot and still cooling. Most materials contract when their temperature falls, and there was no reason to suppose the Earth's rocks were an exception. As the Earth cooled, it must have shrunk, and as it shrank, the solid rocks of its surface would have been drawn together with immense force. This would have crumpled them, and it was this crumpling that produced mountain ranges.

The principal proponent of this theory of mountain formation was the most eminent French geologist of the day, Léonce Élie de Beaumont (1798–1874; see sidebar). Élie de Beaumont studied mountain chains in Europe and North America and discovered that mountains separated by the ocean in fact are related. In all, he studied 21 mountain chains. One such chain consisted of the Carboniferous rocks in the Harz Mountains of Germany, the Vosges and Brittany in France, and Westmorland in northern England, which he found continued in the Allegheny and Ozark Mountains in the eastern and southeastern United States. He also believed that the Pyrenees and European Alps were part of the same mountain system. Clearly all of these mountains had formed at the same time within a single event, and Élie de Beaumont suggested that episodes of mountain building—now called *orogenies*—coincided with gaps in the fossil record that had been identified by the zoologist and paleontologist Cuvier.

Élie de Beaumont reasoned that if the Earth was cooling, its volume must be decreasing as it contracted, leaving the surface unsupported. This produced stresses in the crust that could be released only through crumpling, which occurred in discrete episodes. He described the rocks as being held in the "jaws of a vise." When the crust gives way, Élie de Beaumont maintained, it does so along lines that run around the Earth parallel to great circles. According to this theory, all mountain ranges that lie parallel to a particular great

LÉONCE ÉLIE DE BEAUMONT: THE FRENCH GEOLOGIST WHO DEVELOPED A THEORY TO EXPLAIN MOUNTAIN FORMATION

Jean-Baptiste-Armand-Louis-Léonce Élie de Beaumont was one of the most distinguished French geologists of his generation. Born on September 25, 1798, at Canon, near the town of Caen in the Calvados district of Normandy, he was educated at one of the most prestigious schools in Paris, the Lycée Henri-IV, where he won first prize for mathematics and physics. The school is the equivalent of both a high school and a university, presently with about 1,000 students at each level. The lycée prepares students for the "grand schools," such as the École Polytechnique, which is where Élie de Beaumont continued his studies after graduating from the Henri-IV. He completed his education at another "grand school," the École des Mines, where he studied from 1819 until 1822. That is where he first became interested in geology. In 1823 Élie de Beaumont and Ours-Pierre-Armand Petit-Dufrénoy (1792–1857) were selected to tour the mines and metalworking factories of England and Scotland with André-Jean-François-Marie Brochant de Villiers (1772–1840), professor of geology at the École des Mines.

In 1827 Élie de Beaumont became Brochant de Villiers's assistant, and in 1835 he succeeded him as professor of geology. In 1843 he was made engineer in chief of mines, and in 1847 he became inspector general. He was appointed vice president of the Conseil-Général des Mines in 1861 and in the same year became a grand officer of the Legion of Honor. He was elected to the Academy of Berlin, Académie des Sciences, and Royal Society of London. In 1852 he was made a senator by a presidential decree, and in 1853 he became permanent secretary to the Académie des Sciences.

His theory of the formation of mountain ranges was first published in 1829, and he expanded it in *Revolutions de la surface du globe* (Revolutions of the surface of the globe, 1830) and *Notice sur le système des montagnes* (Account of the mountain system, 1852), published in three volumes. In addition to his work on mountains Élie de Beaumont collaborated with Dufrénoy to prepare a geological map of France, which was published in 1841.

Élie de Beaumont died at Canon on September 21, 1874.

circle must have formed at the same time in a single event, and he claimed that the European mountain ranges formed 12 such groups, all with beds running in the same direction. A great circle is a circle drawn on the surface of a sphere with its center at the center of the sphere. No two great circles can be parallel to each other, but where there are mountain ranges parallel to great circles, according to Élie de Beaumont, the pattern they form links them in a pentagonal (five-sided) network.

The orogenies had caused the extinction of large numbers of species, accounting for the gaps in the fossil record. The idea that single major events had both raised mountain chains and brought about mass extinctions was essentially catastrophist. For a time in the 1850s it became very popular—almost dogma—especially in France, and in 1853 William Hopkins (1793–1866), a distinguished English mathematician and geologist, expounded it in his anniversary address to the Geological Society of London. But eventually the theory lost favor as Lyell's popularization of the rival theory of uniformitarianism attracted increasingly widespread support (see "James Hutton, Plutonism, and Uniformitarianism," pages 144–149).

Meanwhile, Lyell's friend Darwin had discovered the way coral reefs and atolls develop, explaining how the shells of coral polyps could produce limestone strata at high elevations. On September 6, 1838, while Lyell was revising his *Principles of Geology* for a new edition, he wrote in a letter to Darwin:

> I have thrown the chapter on De Beaumont's contemporaneous elevation of parallel mountain chains into one of the Preliminary Essays, where I am arguing against the supposition that nature was formerly parsimonious of time and prodigal of violence. . . . I should like to know, when you next write to me, how far you consider your gradual risings and sinkings of the spaces occupied by coralline and volcanic islands in the Pacific as leaning in favor of the doctrine that many parallel lines of upheaval or depression are formed contemporaneously. If I remember right, some of your lines are by no means parallel to others, although many are so. In one point of view, your grand discovery proves, I think, in the most striking manner, the weight of my principal objection to the argument of De Beaumont. You remember that I denied that he had proved that the Pyrenees were elevated after the cretaceous period, although it is true that the chalk has been carried up to their summits, and lies in inclined beds upon their flanks; for who shall say that the movement was not going on during the cretaceous period? Now in your lines of elevation, there will doubtless be coralline limestone carried upwards, belonging to the same period as the present, so far as the species of corals are concerned. Similar reefs are now growing to those which are upraised, or are rising.

LEOPOLD VON BUCH—AND UPHEAVALS IN THE EARTH

Élie de Beaumont built his theory of global cooling, shrinking, and crumpling partly on evidence of the violence of certain past events that had been accumulating for some time. Much of this evidence had been gathered by Leopold von Buch (1774–1853), the man Humboldt (see "Alexander von Humboldt, Who Recognized That the Earth Changes over Time," pages 117–119) described as the greatest geologist of his time.

Many geologists around the turn of the 19th century were intrigued by the large rocks littering the North German Plain and parts of eastern France that were clearly not derived from the underlying bedrock. No one could imagine how these "erratics" came to be in the locations where they were found. The matter was finally resolved when it was proved that they had been transported by glaciers (see "Ice Ages," pages 122–126).

Some suggested the boulders had once been embedded in vast blocks of ice and had floated on the surface of an ocean that had since disappeared. This was the explanation offered in 1804 by Berlin high school teacher Erhard Georg Friedrich Wrede (1766–1826). Wrede studied the granite erratics in northern Germany and found them quite different from local granite formations but identical to Scandinavian granites. Clearly these boulders had crossed the Baltic Sea. Wrede suggested that within geologically recent times (that is, within the last few million years) the Earth's rotational axis had shifted, and this had lowered sea levels throughout the Northern Hemisphere. If, prior to that shift, the climate had been markedly colder, then the granite blocks might have been transported frozen inside large ice floes that sank as the sea level fell. The blocks he examined lay on low ground not far from the coast, so his idea seemed plausible, but only in respect to the German erratics. It did not explain the boulders found at high elevations in alpine regions farther south, some of which were huge. Nor did it account for the curious way erratics were aligned with the course of the Rhône River.

Von Buch, a fellow Berliner, studied alpine erratics and traveled widely throughout Europe. He observed active volcanoes in Italy, extinct volcanoes in the Auvergne of France, and rock formations in Norway and Lappland. He was the first to report that Sweden, between Frederikshald and Åbo, is slowly but steadily rising, so the sea level is falling along its coast. He agreed with Wrede that the North German granite erratics came from Scandinavia, but he proposed a different mechanism for their transport.

Von Buch began by plotting the location of the German and alpine erratics and their sources. He found that between erratics and their source there was often a region containing no erratics. It was as though the boulders had been thrown, perhaps by some violent explosion. The method of transport could not have been explosive, however, because that would not account for the pattern of their distribution. Nor could huge boulders have been carried high into the mountains on rafts of ice, as Wrede had proposed. In 1815 von Buch visited the Canary Islands, where he studied their volcanoes. Both von Buch and his friend Humboldt had observed that a major volcanic eruption is often preceded by the uplifting of the surrounding land due to the great pressure below ground. This pressure from local underground heating could release lava onto the surface or raise whole areas to form mountain ranges. The force of the uplift could, von Buch thought, hurl boulders through the air as though fired from a cannon. Subsequently the central part of the raised area would collapse, forming a depression.

Europe was divided by the Napoleonic Wars in the early years of the 19th century, making communications slow and unreliable. For this reason von Buch knew nothing of the work in Edinburgh of Sir James Hall (1761–1832). Hall had also studied the alpine erratics and had an idea of how they might have been moved long distances. Hall was a supporter and friend of Hutton (see "James Hutton, Plutonism, and Uniformitarianism," pages 144–149) and believed that eroded crustal rocks are replaced by new rock released in submarine volcanic eruptions. Hutton's uniformitarianism held that the Earth's surface had been formed and shaped by processes that were still active—that "the present is the key to the past," as Lyell put it—but Hall disagreed with Hutton in maintaining that it did not require those processes to be gradual, and he had a historic event to demonstrate his point. At 9:45 A.M. on Sunday, November 1, 1755, an earthquake had destroyed most of Lisbon and several other Portuguese cities. Shortly after the earthquake a series of tsunamis swept over Lisbon harbor and submerged much of the low-lying parts of the town. Scientists knew that submarine earthquakes could trigger tsunamis, but Hall suggested that the tsunamis recorded in history, including those that struck

Lisbon, were minor events compared with what might be possible. If a major volcanic eruption abruptly lifted the ocean floor, the resulting tsunami might be sufficient to carry water far inland, traveling with great energy. Hall tested this idea experimentally by firing underwater explosions and observing the effect when the shock wave reached the shore. If such an event occurred at a time when large boulders were frozen inside ice, making them buoyant, the giant tsunami might hurl them long distances.

The French geologist Dieudonné-Sylvain-Guy-Tancrède de Dolomieu (1750–1801), usually known as Déodat de Dolomieu, had suggested as early as the 1790s that "megatsunamis" had played a key role in the formation of rocks. He calculated that the transport of entire beds of rock and subsequent scouring of the land surface would have required tsunamis up to 5,250 feet (1,600 m) high.

Hall and von Buch had reached similar conclusions. André-Jean-François-Marie Brochant de Villiers (1772–1840), professor of geology at the École des Mines in Paris, was greatly impressed by von Buch's work. Élie de Beaumont heard of it initially from Brochant de Villiers and saw it as evidence of the crumpling caused by the Earth's contraction.

Christian Leopold, Baron von Buch was born on April 26, 1774, at Stolpe an der Oder, Pomerania, a region of northern Europe that was then part of Prussia and is now mainly in Poland. In 1790 von Buch enrolled at the Freiberg Mining Academy, where he studied under Abraham Gottlob Werner (see "Abraham Gottlob Werner—and the Classification of Rocks," pages 114–117), graduating in 1793. While there he and Humboldt, a fellow student, became friends. Von Buch completed his education at the Universities of Halle and Göttingen, in Germany. He obtained a position as an inspector of mines in 1796, but being independently wealthy, he soon resigned in order to devote himself wholly to his travels and research. He visited the Alps in 1797, and in 1798 he went to Italy, where his examination of the rocks of Mount Vesuvius raised doubts about Werner's theory that all rocks are formed by sedimentary processes on the ocean floor. His visit to the Auvergne in 1802 confirmed for him that volcanism had once been widespread in Europe and that massive upheavals of the Earth had raised the Alps. He traveled to Scandinavia in 1806, where he identified the source of the erratic rocks found in northern Germany. Von Buch died in Berlin on March 4, 1853.

Von Buch left Freiberg a convinced Wernerian, but both he and Humboldt were unable to ignore compelling evidence that mountains rise as a result of uplift, not sequential sedimentation. They did not reject sedimentation as a mechanism, but they did reject Werner's idea of a retreating ocean.

CONSTANT PRÉVOST—AND THE SHRINKING EARTH

Von Buch had his opponents, among them Constant Prévost (1787–1856), professor of geology at the Athenaeum, an academic institution in Paris in the early 19th century. Prévost never wavered in his uniformitarianism, rejecting entirely the idea of catastrophic events of the kind implied by von Buch's upheavals. He supported Élie de Beaumont's theory of a contracting Earth, comparing the process to crinkling of its skin as a stored apple slowly dries.

Louis-Constant Prévost was born in Paris on June 4, 1787. He studied in Paris, graduated in letters and sciences in 1811, and began studying medicine and anatomy, but influenced by Brongniart, he later turned to geology. One of the founders of the Societé Géologique de France, Prévost was professor of geology at the Athenaeum from 1821 to 1829, and in 1831 he was appointed an assistant professor and later honorary professor of geology at the Sorbonne. He died in Paris on August 14, 1856.

Cuvier and Brongniart had collaborated on a study of the rocks of the Paris Basin, and in 1808, shortly after they had presented their joint paper at the Institut National, Brongniart departed on a field trip to the south of France, taking his student Prévost with him. Officially they were seeking sources of kaolin (china clay) for the porcelain works at Sèvres, but this allowed them to visit the Pyrenees and then the Massif Central, where they examined limestone formations and the fossils these contained.

Prévost was also an associate of the engineer and inventor Philippe de Girard (1775–1845). In 1810 the French government had organized a competition with a prize of 1 million francs for whoever could invent the best machine for spinning flax. Girard patented devices for spinning wet and dry flax but failed to win the prize. Nevertheless, in 1815 the Austrian government invited him to establish a spinning mill near Vienna. Prévost accompanied Girard to Austria, where he spent the years from 1816 to 1819 making a study of the Secondary

(Mesozoic) and Tertiary (Cenozoic) rocks of the Vienna Basin. He examined the fossils but interpreted them in a new way. The custom was to look for fossils that were characteristic of a stratum because they were striking and not found in other formations. Instead Prévost looked for assemblages of fossils and categorized their species compositions. He could then use assemblages with that general composition to identify rock formations, even if those formations contained no individual species found nowhere else. This was a major advance in stratigraphy.

In 1821 Prévost published an essay on the geology of parts of Normandy where he found Secondary (Mesozoic) rocks similar to formations found in southern England. He collaborated with Lyell on this project and supported Lyell's uniformitarianism, maintaining that volcanic peaks grow gradually, through the accumulation of lava from successive eruptions. In *De la chronologie des terrains et du synchronisme des formations* (On the chronology of terranes and the synchrony of formations) published in 1845, Prévost described the way successive stages of *igneous* intrusion and sedimentary deposition occurred at the same time across large areas.

HORACE-BÉNÉDICT DE SAUSSURE— AND THE STORY OF THE ALPS

The European Alps are scenically magnificent and have been attracting sightseers, climbers, and winter sports enthusiasts for more than two centuries. The illustration reveals the beauty of this mountain range, located in the heart of a densely populated continent. The first visitors to climb the mountains were not there for the mountaineering, however, and far less for the winter sports. They were scientists studying the weather and the way mountain ranges form.

Mont Blanc, towering to 15,771 feet (4,810 m) above sea level, is the highest mountain in Europe. Located on the border between France and Italy, it is best reached from the French town of Chamonix. The first person to climb the mountain, on August 8, 1786, was a local Italian doctor, Michel Gabriel Paccard (1757–1827), accompanied by Jacques Balmat (1762–1834) from Chamonix. A year later, on August 3, 1787, Balmat climbed the mountain again, this time as guide to Horace-Bénédict de Saussure (1740–99), a Swiss scientist who is credited with having started the fashion for visiting and

climbing in the Alps, although travelers had been visiting the area for some years to admire the scenery and wonder at the glaciers. On his first visit to Chamonix in 1760, Saussure had offered a prize to the first person to reach the summit of Mont Blanc, and he made an unsuccessful attempt on the mountain himself in 1785. He made his successful ascent in the company of 18 porters and guides, in addition to Balmat.

Saussure spent a great deal of time in the Alps over the years, and he visited Mont Blanc several times. The first volume of his most famous book, *Voyages dans les Alpes* (Travels in the Alps), had been published in 1779, and three more volumes appeared between then

This view of Grindel-wald, Switzerland, is from Männlichen looking down into the Lauterb-runnen Valley. *(Stephen Studd/Stone/Getty)*

and 1796. In *Voyages* Saussure described seven of his journeys into the mountains. His interest was not in sport, however, but science. He studied alpine plants—the genus *Saussurea* is named in his honor—and meteorological conditions. Above all, though, Saussure was a geologist and one of the first to adopt the term *geology* (*géologie,* in French) introduced in 1778 by another native of Geneva, Jean-André de Luc (1727–1817), as an alternative to the older *geognosy.*

When his party stood at the summit of Mont Blanc, they raised a flag as a signal to Saussure's wife and her sister, who were watching through a telescope from Chamonix. Then Saussure looked out on the surrounding peaks. He saw the way they were located in relation to each other and their structure. While he was doing this, other members of the party were erecting a tent and table and setting out the instruments with which Saussure measured the air pressure and temperature and confirmed that Mont Blanc was higher than any of the other mountains around it. He later converted the pressure readings to altitudes, calculating that the peak was 15,683 feet (4,780 m) above sea level.

Saussure aimed to discover a general "true theory of the Earth," a comprehensive account of its formation and the steps by which it had reached its present state. His love of mountains was central to this ambition, because he believed it was there, where most rocks are exposed, that the evidence he needed would be found. He studied fossils and rock strata and also the ways strata were deformed. On his journey from Geneva to Chamonix he passed a place where sedimentary rocks were exposed in a huge fold. He made meticulous field notes of everything he observed and warned the would-be geologist (*géologue* was the term he used) that the life was hard, involving long, tiring, and hazardous journeys and many hours indoors spent quietly examining specimens. Saussure did much to change geology into a rigorous scientific discipline.

Saussure was born at Conches, near Geneva, on February 17, 1740. His education began when he enrolled at the public school in Geneva in 1746, at the age of six. In 1754 he entered the Geneva Academy (University), graduating in 1759, when he was 19. His dissertation was on the physics of fire. In 1761 he applied unsuccessfully for the position of professor of mathematics at the Geneva Academy, and the following year he applied again, this time successfully, for the professorship of philosophy and natural sciences. In 1772 Saussure was elected a fellow of the Royal Society of London, and in the same year he founded the Society for the Advancement of the Arts in Geneva. By this time Saussure was a well-established and respected physicist. He was rector of the academy in 1774–75.

Saussure spent much time traveling, in addition to his visits to the Alps. He married Albertine Boissier in May 1765 and in 1768–69, accompanied by his wife and her sister, he visited Paris, the Netherlands, and England. He visited Italy in 1771, and in the autumn of 1772 he toured Italy, this time in the company of his wife and their six-year-old daughter. He climbed Mount Etna, called at Rome and had an audience with Pope Clement XIV, went to Rimini and Venice, and returned to Switzerland over the Brenner Pass.

The late 18th century was a time of political unrest in France and also in Switzerland, and in 1776 Saussure drew up plans for the reform of city institutions in Geneva. He was arrested during disturbances in 1782 and spent two days in prison, and in July of that year he was besieged for several days in his own home, suspected of harboring armed men and concealing weapons.

He resigned from the Geneva Academy in 1787 and moved for a time to the south of France, where the climate was more congenial and where he could collect measurements of sea-level atmospheric pressure to compare with those he had made in the Alps. The Terror that followed the French Revolution spread to Geneva, and there was a prodemocracy revolution in 1792. During a brief period of calm, in 1793, Saussure, still living in France, was appointed to the commission charged with drafting a constitution. The constitution failed, however, and in 1794 the Terror returned.

Throughout all of this Saussure continued with his fieldwork and writing, but his health was failing. In 1794 he suffered a stroke that brought his field excursions to an end. He was also in financial difficulties that compelled him to return to the family home at Conches. He received offers of help from abroad, and Thomas Jefferson considered offering him a position at the newly formed University of Charlottesville, in Virginia, but it was not to be.

Saussure never did discover his "true theory of the Earth," although he compiled a table of contents for it and wrote the first three chapters and part of the fourth. On December 21, 1796, he suffered a second, more massive stroke from which he never recovered. Saussure died at Conches on the morning of January 22, 1799. He is buried in the Plainpalais cemetery in Geneva.

NEPTUNISTS VERSUS PLUTONISTS

Neptune was the Roman god of the sea. Pluto is often described as the god of the underworld. In the 18th century Neptune and Pluto were the names chosen to symbolize two rival theories describing the origin and formation of the Earth. Contrasting the Neptunists and Plutonists makes it sound as though the two gods were at war, but if there was conflict, it was conducted very politely. Furthermore, linking the Neptunists to the idea of a universal flood makes them sound old-fashioned and the Plutonists, whose views in the end largely prevailed, appear more modern and enlightened by contrast, but this, too, is a great oversimplification.

The late 17th and 18th centuries were the time of the Enlightenment, when all over Europe thinkers were investigating and debating the way the world and the universe function. These philosophers took nothing for granted and accepted nothing simply on the ground that it was a long-held belief. One of the problems exercising them concerned the way the Earth had acquired its present surface, and they were prepared to challenge tradition.

Scientists knew that sedimentary rocks formed by the compression of sediments that accumulated on the seabed, but sedimentary rocks were found everywhere, often far from the coast. How did they come to be there? Hooke had suggested that the Earth's surface changed through time, but his idea found little support. In 1691 the German philosopher and mathematician Gottfried Wilhelm Leibniz (1646–1716) wrote a work called *Protogaea* in which he explored the idea that the world's oceans had once been much deeper than they were at present and that they had steadily retreated over a long period, leaving behind the sedimentary rocks that had formerly lain on the ocean floor. *Protogaea* was not published in full until 1749, however, and in 1748 the idea was published in a much more radical form in a book by French diplomat, traveler, and writer Benoît de Maillet (1656–1738). De Maillet wrote his book toward the end of the 17th century, at about the time Burnet's *Sacred Theory of the Earth* was published, and it circulated fairly secretly for many years before it was finally published, 10 years after its author's death. It was a scandalous book that contradicted the biblical account of creation, and in order to make it acceptable to the church de Maillet wrote it in the form of a fictional discussion between a French missionary and an Indian philosopher called Telliamed—*de Maillet* written backward. *Telliamed* was also the title of the book (in full, *Telliamed or Conversations Between an Indian Philosopher and a French Missionary on the Diminution of the Sea*). Obviously, no one could expect Telliamed to be familiar with Christian teaching, and the missionary merely recorded what Telliamed said to him.

Telliamed recounted what he had been told by his own grandfather, who had observed that the coastline near to his village had retreated during his lifetime, so the sea was now farther away than it used to be. De Maillet based this assertion on detailed records the Egyptians had kept of the annual Nile floods. These appeared to show that sea level had fallen during historic times. Telliamed's grandfather then turned his attention to the hills near his village and to the seashells contained in their rocks, which proved that the hills had once lain beneath the sea. The rock strata were of different colors, and the shells in one stratum were different from those in another. This showed, said Telliamed, that the land must have been inundated several times, but there had been no inundation in recent times. He proposed that the hills had been made from material piled up by ocean currents. As the sea retreated, animals formed on the land nearest to the surface. Later, as the sea level continued to fall, the marine animals turned into land animals. All of these processes occurred over a very long expanse of time. It meant the world was immensely ancient. De Maillet suggested it was billions of years old, although Abbé J. B. le Mascrier, who was responsible for its 1748 publication, altered *billions* to *millions* to make the book slightly more acceptable.

There were some English geologists who linked Neptunist theory to the biblical flood, but very few of the French and German Neptunists did so. They sought merely to explain the origin of sedimentary rocks. Werner's interest in Neptunism arose from his chemical theory about the formation of minerals (see "Abraham Gottlob Werner—and the Classification of Rocks," pages 114–117). If, as he believed, minerals crystallize out of a saturated aqueous solution, clearly there must once have been abundant water everywhere—one or more universal floods.

Neptunism was mistaken, but not because it was derived from attempts to support the Genesis flood story. It was mistaken because it predated the theory of plate tectonics (see chapter 8), which provides a mechanism by which seafloor sediments form sedimentary rocks that are later uplifted high above the sea without recourse to universal floods.

If the neptunists were not reactionaries clinging to outdated ideas, neither were all of Pluto's supporters modern visionaries. Hutton (see "James Hutton, Plutonism, and Uniformitarianism," pages 144–149), the principal proponent of the idea that new rocks are formed from magma and old rocks erode in an endless cycle of growth and decay, aimed to describe a self-sustaining world created by a loving God. He maintained that the Earth was infinitely old and would last forever. In fact, in his view Earth is perfect, as would be expected of the creation of an all-wise, all-knowing God. His vision was essentially religious, and although Hutton suggested the kind of evidence that would support his theory, he wrote it before making any detailed study of volcanism and before seeing his first *unconformity* in 1787, where horizontal sedimentary strata lie on top of strata that have been tilted until they are almost vertical. Hutton concluded, correctly, that the horizontal strata must have been deposited on top of the older tilted strata. The illustration shows how an unconformity can occur. When, in 1788, Hutton published his ideas in the *Transactions* of the Royal Society of Edinburgh and then in 1795 as a two-volume book, he found himself attacked from two sides. From one direction came those who were shocked at what they saw as atheism in his total repudiation of the Genesis story, and from the other side came attacks from those who objected to the idea of the Earth having been perfectly fashioned by God. When Playfair published his *Illustrations of the Huttonian Theory of the Earth* in 1802, he acknowledged that Hutton's theory implied there could be no trace of the event of creation but rejected any suggestions that Hutton ever intended to deny the creation.

As expounded by Playfair, Hutton's theory was compelling and gradually most of the Neptunists came round to it (see, for example, "Alexander von Humboldt, Who Recognized That the Earth Changes over Time," pages 117–119). They accepted the importance of volcanism in rock formation, but they rejected extreme interpretations of Hutton's idea that processes occur gradually and continually so the Earth does not develop and change. In a word, Hutton implied that the Earth has no history but has always been very much as it is now and will always remain that way. The "reformed Neptunists" maintained that the present state of the Earth's surface

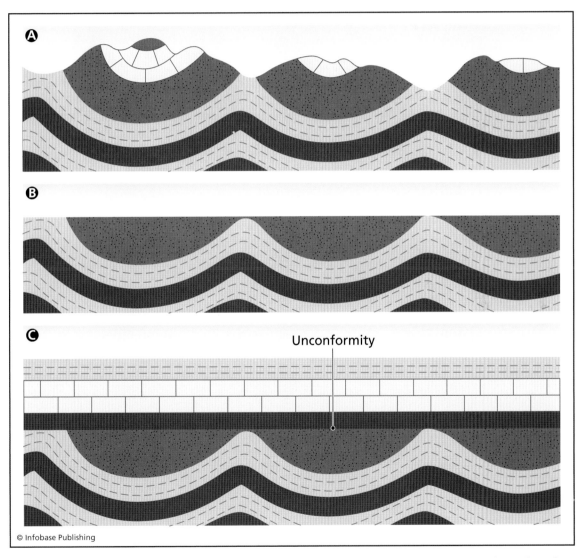

An unconformity, of the type Hutton proposed and then discovered, develops in stages. The rock surface is uplifted (A), exposing it to weathering. Weathering levels the surface (B). New sediments accumulate on top of the old surface (C). The boundary between the older and newer strata is an unconformity.

results from a sequence of events, a history that had a beginning with the Earth's formation.

JAMES DWIGHT DANA— AND THE PERMANENT CONTINENTS

Prévost believed that as the Earth cooled, it would shrink, and as it shrank, its surface would wrinkle, like the skin of a drying apple (see "Constant Prévost—and the Shrinking Earth," pages 161–162). This mechanism for producing mountains is fundamentally different from the ideas of the Plutonists, because it proposes not that mountains are thrust upward (by whatever means), so the forces act vertically, but that the forces act horizontally, by lateral compression. James Dwight Dana (1818–95), the most distinguished American geologist of his generation, shared Prévost's opinion, and in the 1840s he used it to explain the geologically complex Appalachian Mountains of the eastern United States. The photograph of the Great Smoky Mountains, which are part of the Appalachian system, gives an idea of the difficulties any geologist might face in trying to understand how they formed.

If the Earth's surface contracts, lateral compression will raise some areas and lower others, crumpling the rocks like a tablecloth that is pushed from either side toward the center. In Dana's view this picture is complicated by the fact that the surface cooled unevenly, the temperature falling faster in some parts than in others. At the same time uneven cooling below the surface produced extensive depressions, where surface rocks sank with the contracting material beneath it, like a cake that falls in the center during baking. Dana believed that the portion of the Earth's crust separating North and South America from Europe and Africa sank in this way to form a huge depression. Water always flows toward the lowest point, and it filled the depression to become the Atlantic Ocean. The Atlantic Basin and the continents on either side formed very early in the history of the Earth, and their essential features have remained largely unaltered ever since. Continents and oceans, in Dana's view, are permanent structures.

Cooling continued, however, and that is when the lateral compression occurred. Erosion is also a continual process. Dana argued that the continental rocks—exposed to wind, water, and seasonal heating

and cooling—eroded, producing particulate material that was carried to the sea, where it accumulated as sediment that was hardened to become sedimentary rock. Lateral compression then crumpled the sedimentary rocks, producing mountain ranges along the coasts of continents. The Appalachians were made from sediments eroded from the interior of the continent, hardened into rock, and later folded to their present shapes by lateral compression.

Dana's description sounded plausible, but at first it overlooked the fact that sediments containing marine fossils are also found deep in the interior of continents. This did not square with the idea of eroded material being transported to the coast, where sedimentary rocks

The Appalachian Mountains are geologically complex. This view, photographed in about 1980, is of the Great Smoky Mountains, a part of the Appalachians occupying 800 square miles (2,070 km²) on the border between North Carolina and Tennessee. *(Getty Images)*

form on the seabed. The solution to this difficulty might be that at times the movements of the rocks due to contraction might allow the sea to inundate low-lying areas, even areas far from the coast. These inland seas would be shallow, but sedimentary rocks could form beneath them. Alternatively, perhaps the continental rocks rested on top of denser and therefore heavier rocks that extended beyond the coast and formed the ocean floor.

Today geologists know that the rocks making up continental crust are less dense than those composing oceanic crust, and that the continental crust is up to 37 miles (60 km) thick, whereas the thickness of oceanic crust averages three miles (5 km). They also know that lateral compression is real and that it raises mountain chains, as well as that the Appalachians formed a very long time ago—about 420 million years ago, in fact. So Dana was correct on several points, but his belief in the permanency of continents and ocean basins was mistaken. Geologists now believe that the Appalachians, as well as the mountains of Greenland, the Highlands of Scotland, and Scandinavia all formed at the same time when an ancient ocean, called the Iapetus, disappeared because continents were moving toward each other. The illustration shows the Iapetus Ocean and the lands surrounding it as these may have been arranged about 480 million years ago.

Dana was born on February 12, 1813 in Utica, New York, where his father owned a hardware store. His mother was deeply religious and strongly influenced him. The eldest of four children, Dana became adept at using tools and was artistic and musical (he played the piano and guitar). He also enjoyed walks in the countryside, collecting rocks, plants, and insects. His formal education commenced at Utica High School, where Fay Edgerton, one of the teachers, encouraged his interest in science. In 1830 he entered Yale College, where he studied under the chemist Benjamin Silliman (1779–1864). Dana studied several disciplines, graduated in 1833, and for the next two years he taught mathematics to navy midshipmen, serving on board the USS *Delaware* and USS *United States*. He sailed to the Mediterranean, where he saw an eruption of Mount Vesuvius. His first scientific paper was a description of the eruption, published in the *American Journal of Science,* which was edited by Silliman. In 1836 and 1837 he worked as an assistant to Silliman, developing a system for classifying minerals based on their chemistry and crystallography. He published this in 1837 as *System of Mineralogy,* and it was very successful. He followed it with *Manual of Mineralogy* in 1848 and *Manual of Geology* in 1863.

Dana obtained a position as mineralogist and geologist to the United States Exploring Expedition (1838–42), charged with charting Pacific islands and visiting Antarctica. The expedition, involving

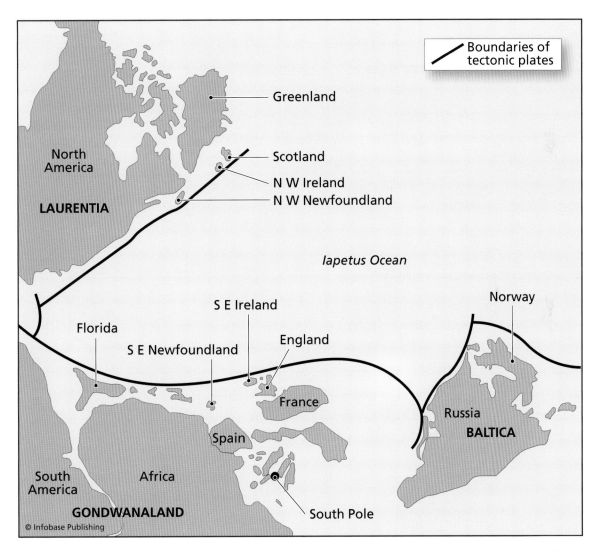

The Iapetus Ocean as it was about 480 million years ago, during the Ordovician period. There were three major continents. Laurentia consisted of modern North America, with Greenland, northern Scotland, northwestern Ireland, and northwestern Newfoundland as islands. Gondwanaland comprised South America and Africa, with Florida and most of Europe as islands. Baltica included Scandinavia and Russia. The heavy lines represent the boundaries of tectonic plates.

six ships under the command of Acting Captain Charles Wilkes (1798–1877), was the first oceanographic expedition to be funded by the U.S. Navy. When it ended, Dana spent the following 13 years preparing his reports.

In 1844 Dana married Henrietta Frances Silliman, Benjamin Silliman's daughter, and the couple settled in New Haven, Connecticut. In 1850 he was appointed Silliman's successor as Silliman Professor of Natural History and Geology at Yale, a post he held until 1892, and in 1846 he became chief editor of the *American Journal of Science and Arts* and a frequent contributor to it.

Dana's theory of permanent continents and the mountain-building consequence of compression due to shrinking was the most widely supported explanation of how mountain chains form until the advent of the plate tectonic theory, long after his death. He believed that the processes of volcanism, erosion, subsidence of volcanic chains, and compression led to the growth of continents. As continents became larger, the climate became harsher and living organisms became more highly complex. Deeply religious throughout his life, Dana saw this as evidence that what he called the "Power Above Nature" had prepared the Earth for humans, who are the purpose and end point of history.

His health had never been robust, and in his later years Dana moved to Hawaii, where the climate was more congenial. He was still working at revising the text for a book he was writing on volcanoes, writing scientific papers, and answering letters until shortly before his death on April 14, 1895.

EDUARD SUESS, COLLIDING ROCK MASSES, AND MOVING CONTINENTS

Influential as Dana was, his concept of permanent continents had an opponent in Eduard Suess (1831–1914), one of Europe's most eminent geologists. Suess studied the Alps, which had a structure different from that of the Appalachians and called for a different explanation.

Dana's proposition that the Atlantic Ocean filled a basin caused by the subsidence of the crust is known as the *geosyncline* theory. During the 19th and early 20th centuries this was widely accepted, and the theory became very detailed, with an elaborate terminology, most of which has now been abandoned. Suess accepted the geosyncline theory in general and agreed that the Earth was growing cooler, but he differed from Dana over the way these processes worked.

Suess used evidence from his studies of alpine structures to propose that although contraction produced lateral compression, there was an alternative to folding as a way to relieve the pressure this created. What had happened in the Alps was that rock strata had fractured and the rocks on either side of the fractures had moved together, with rocks on one side riding over the rocks on the other side. This was overthrusting, and Suess maintained that the Alps had formed fairly recently—in geologic terms—as the result of a massive northward thrusting movement. This event and others like it were not catastrophic in the sense of being sudden and violent, but they did occur as discrete episodes of change.

In part Suess was seeking to resolve a dilemma. Most geologists agreed with the uniformitarian view that the processes that had shaped the surface of the Earth were still active. These processes were local and occurred at different times and rates in different parts of the world. The fossil record, on the other hand, showed organisms appearing and disappearing so often that fossil assemblages could be used to date rock strata from all over the world. The fossils suggested worldwide, catastrophic extinctions and the relatively sudden emergence of new groups of organisms. Suess was able to show that sea levels had often changed simultaneously all over the world and that the composition of fossil assemblages was linked to these changes. Sea levels changed due to the episodic sinking of the ocean floor due to contraction. He coined the term *eustasy* to describe such changes in sea level. Eustasy explained why the use of fossils to date strata was so successful.

Suess was born on August 20, 1831, in London, England, the son of a merchant from Saxony, now part of Germany. When Eduard Suess was three, his family moved to Prague, then the capital of Bohemia, and when he was 14, the family moved to Vienna. Suess is always described as being Austrian. He was educated in Vienna and at the University of Prague, returning to Vienna after he graduated. From 1852 until 1856 he worked as an assistant in the Hofmuseum (now the Natural History Museum) in Vienna. While there he published papers on ammonites and brachiopods.

Suess published his first work on the Alps, a short book called *Die Entstehung der Alpen* (The origin of the Alps), in 1857. In it he argued

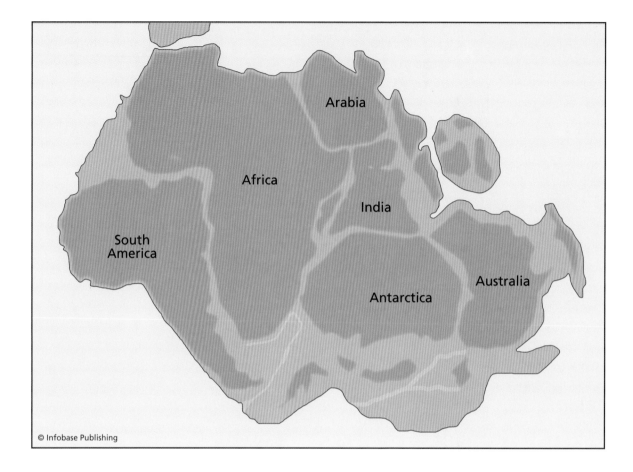

© Infobase Publishing

The supercontinent of Gondwanaland (Gondwana) as it is now thought to have appeared during the Carboniferous period (359.2–299 million years ago)

that the Alps resulted from horizontal movements of the Earth's crust, producing folding and overthrusting, and that volcanism was the consequence of mountain building rather than its cause. He considered the possibility that North Africa and Europe might once have been joined by a land bridge.

His most influential work, *Das Antlitz der Erde* (*The Face of the Earth*), was published in four volumes between 1883 and 1909. In it Suess described his theories of the way the Earth formed and has developed by cooling and shrinking, and at the same time how the rate of change has steadily slowed down. He maintained that erosion shifted material from the interior of continents to the coasts, where it accumulated as sediments filling basins. The basins filled, raising the sea level and allowing the sea to invade low-lying regions of the continents. From time to time a section of crust would collapse, forming a new geosyncline into which water drained from the inland seas.

In the course of this exposition Suess suggested that at one time South America, Africa, India, and Australia had been joined together in a single supercontinent that he called Gondwanaland, which had broken apart later, when its interior collapsed, producing the present continents. Later geologists found that other parts of southern Asia, Madagascar, Antarctica, and Florida also formed part of Gondwanaland (at present usually called Gondwana). The map shows Gondwanaland as modern geologists believe it appeared during the Carboniferous period. Suess also proposed the existence of the Tethys Ocean. Suess was the first person to recognize that major rift valleys, such as the East African rift system, are caused by extension of the lithosphere.

In 1856 Suess was appointed professor of paleontology at the University of Vienna, and in 1861 he was made professor of geology. By the end of his life Suess had done much to pave the way for modern paleogeography—the geography of the remote past—and the structure and movement of the Earth's crust. He died in Vienna on April 26, 1914.

Drifting Continents and Plate Tectonics

Throughout the period from the early 18th to middle 20th centuries, geologists sought to reconstruct the history of the solid Earth to account for the processes and steps by which it formed and assumed its present structure and features. This required that they recognize that the surface of the Earth changes through time and, therefore, that the planet has a history, with a beginning followed by a sequence of events. The geologists discovered much, but they still lacked one final concept, a key that would allow everything else to fall into place. The work of Eduard Suess, in the second half of the 19th century, came close to finding that key, but it was left to scientists of the 20th century to complete the picture.

This chapter tells of the discovery of the theory of plate tectonics, the unifying idea that has come to form the basis of modern geology. This part of the story begins with the realization that although the Earth might be cooling, any resulting contraction would be far too small to raise mountain ranges, so some other mechanism must be at work. Various alternatives were proposed, and then eventually, backed by a vast armory of observational evidence, a German meteorologist made the seemingly outrageous suggestion that over the course of history entire continents have wandered about the Earth's surface. He was saying that the map of the world with which everyone is familiar today would be hopeless at depicting the world that existed hundreds of millions of years ago and that the present world map does not depict the world that will exist millions of years from now.

As happens with many important scientific ideas, this one was denounced as preposterous and consigned to the intellectual scrap heap, where it remained until a mechanism was discovered by which it might work. Years later, clear evidence was found to show that continents have indeed moved and to this day are still moving, at an average rate, though with wide variation, of about one inch (2.5 cm) a year. Continents move at about the same speed as toenails grow.

OSMOND FISHER—AND FLOATING CONTINENTS

Supporters of James Dana held that the continents are permanent. Supporters of Eduard Suess believed there had once existed a continent they called Gondwanaland that had broken apart, showing that continents are impermanent. On this the two groups held divergent views, but both agreed that the Earth was steadily growing cooler and that mountains formed as a result of the contraction of surface rocks, although they disagreed on precisely how this happened. Physicists calculated, however, that contraction through cooling could not generate enough force to raise mountain chains. One geophysicist who made this point very strongly was the Reverend Osmond Fisher (1817–1914).

Fisher was born on November 17, 1817, in Osmington, Dorset, in the south of England. He was a Church of England clergyman for all of his adult life but devoted his free time to geology, studying the *geomorphology* of Norfolk, and the stratigraphy and fossils of his native Dorset. He died in Huntingdon on July 12, 1914.

Fisher described his most important work in an appendix to the second edition (1892) of his book *The Physics of the Earth's Crust*, first published in 1881. Geysers, hot springs, and volcanoes release boiling water, hot rock, ash, mud, and lava from locations scattered all over the world. This suggests that the material below the ground surface is at a higher temperature than the surface rocks. Miners had long known that the temperature increases with increasing depth below ground, even in the permafrost regions of Siberia. Although the rate of temperature change varies from place to place, it was clear by the late 19th century that if the temperature continued to increase with depth at an average rate, the interior of the Earth would be hot enough to melt rock below a depth of about 20 miles (32 km). But there were problems with this calculation, and other scientists had calculated that either the whole of the interior was solid or, if it was partly liquid, the solid crust must be approximately 1,000 miles (1,600 km) thick. Sir George Howard Darwin (1845–1912) and Lord Kelvin had shown experimentally that a rotating oblate spheroid behaves under stress as though it were solid even if it was liquid, so effectively the Earth must be solid throughout. There was also the question of the tides. If the Earth's interior were liquid, the tidal attraction of the Moon and Sun would make the surface rise and fall; in that case, the land would rise together with the ocean and there would be no visible tides along coasts.

Fisher replied to all of these objections, and others, by arguing that the molten interior of the Earth is able to expand and contract and is in constant motion. He suggested several experiments, which were performed by others. From these Fisher estimated the average density of the crust to be 2.68 times that of water and the density of the molten interior to be 2.96 times that of water. Fisher then calculated the average thickness of the crust at sea level to be 18 miles (29 km). If the molten interior were not in motion, the rate of cooling would have produced a crust 18 miles (29 km) thick by the time the Earth was about 8 million years old, but geologists agreed that the rocks of the crust must be at least 100 million years old. This could be possible, Fisher maintained, only if the interior were moving and convection currents were constantly transporting heat from near the center of the Earth to the region immediately beneath the crust, thereby preventing the crust from growing thicker.

Years ahead of his time in arguing so strongly for tidal and convective movements in the Earth's interior, Fisher had proposed a mechanism by which continents might be transported across the planet's surface. But no one at the time was interested in the idea of moving continents, and Fisher's work was largely ignored, although it found one enthusiastic supporter in Alfred Russel Wallace (1823–1913), the naturalist who independently discovered evolution by natural selection at the same time as Darwin. Fisher's ideas were largely forgotten, but the steps needed to revive interest in them were already beginning, even as he was preparing his appendix to *The Physics of the Earth*.

CLARENCE DUTTON—AND ISOSTASY

Earthquakes occur when rocks below the surface break and move abruptly as a result of accumulated stress, causing movement at the surface. The sudden movement releases shock waves that travel through the Earth along different paths and at different speeds, causing the rocks to oscillate about a fixed point as the waves pass. Some waves travel close to the surface, others travel at a deeper level, and some pass through the Earth's core. The study of earthquakes is called seismology, and modern seismology relies on recording the waves as they pass and examining them as traces made by a pen on a chart. As well as revealing the actual location of the earthquake, known as the hypocenter, or focus, studying the waves provides information on the material through which they have passed.

The American geologist Clarence Dutton (1841–1912) was one of the founders of seismology. In 1886 a large earthquake shook Charleston, South Carolina, and in 1890 Dutton wrote a monograph on what caused it. This was one of the first thorough scientific reports on an earthquake. Dutton also studied volcanism in Hawaii, Oregon, and California from 1885 to 1888, and he was an authority on the history of the Grand Canyon.

Dutton made what was probably his most important contribution to geology in 1889 when he described the way continents float like rafts on the Earth's mantle, finding their own level as their weight changes. He named this process of self-adjustment *isostasy*, literally "equal station," from the Greek *isos*, meaning "equal," and *stasis*, meaning "station." Dutton realized that the weight of a continent would depress the underlying rocks, causing the continent to sink, just as a ship sits lower in the water when its holds are full of a heavy cargo than it does when it has unloaded the goods it carried. Dutton then went on to suppose what would happen as sediment accumulated on a continental surface and was removed from it by erosion. The accumulation of sediment would increase the continent's weight, and it would sink lower. Then, as erosion removed the sediment, its weight would decrease, and it would rise. Isostasy is a process of adjustment that restores gravitational equilibrium after mass has shifted from one place to another.

It was a powerful idea that other geologists developed. What neither they nor Dutton knew was that the image of continents floating on a denser material left room for continents to move, if the underlying medium was sufficiently plastic to permit it.

Clarence Edward Dutton was born on May 15, 1841, in Wallingford, Connecticut. His father was a shoemaker and for many years the town postmaster. Clarence Dutton attended school in Ellington and qualified for entry to Yale University when he was only 14 but waited a year, enrolling in 1856 and graduating in 1860 when he was just 19. He continued with postgraduate studies until 1862. At first he intended to become a minister, but then he was drawn to literature and also to mathematics and chemistry.

In 1862 Dutton enlisted as a second lieutenant in the 21st Connecticut Volunteers and was quickly promoted to adjutant, with the rank of first lieutenant. He fought in several Civil War battles and was wounded at Fredericksburg. He was promoted to captain in 1863, the year when he passed the necessary examination for admission to the Ordnance Corps of the regular army, joining the corps in January 1864 as second lieutenant of ordnance. He spent the next 11 years at arsenals in Troy, New York; Frankfort, Kentucky; and Washington, D.C. During this time he had opportunities to develop his scientific interests.

By 1875 Dutton held the rank of captain of ordnance, and in 1875 the Ordnance Corps detached him to the Interior Department to take part in the U.S. Geographical and Geological Survey of the Rocky Mountain region accompanying the soldier, explorer, and geologist John Wesley Powell (1834–1902), who became a firm friend. In subsequent years Dutton published reports on the high plateau of Utah and the Tertiary (Paleogene and Neogene) history of the Grand Canyon. His two reports on the Grand Canyon, published in 1882, were among the first to be published by the U.S. Geological Survey, which had been formed in 1879. Dutton was elected to the National Academy of Sciences in 1884.

His studies of volcanism led Dutton to an interesting conclusion. In "Volcanoes and Radioactivity," a paper published in 1906, he proposed that the decay of radioactive elements beneath the Earth's crust releases the heat that melts the rocks to form lava. The weight of the overlying solid rock then bears down on the molten rock, forcing it to the surface through fissures.

He returned to the Ordnance Corps in 1890 at his own request and was appointed commander of the arsenal at San Antonio, Texas, with the rank of major, and later ordnance officer for Texas. Dutton retired in 1901 but maintained an active interest in geology. He went to live with his son, Clarence Edward, Jr., in Englewood, New Jersey, where he died on January 4, 1912.

THOMAS CHAMBERLIN—AND THE CYCLE OF EROSION

By the end of the 19th century geologists had abandoned the idea of the contracting Earth but still believed continents to be permanently in their present positions. Clearly the land surface underwent change through erosion and sedimentation and Dutton's concept of isostasy offered an explanation of another step in the evolution of continents. It was the American geologist and cosmologist Thomas C. Chamberlin (1843–1928) who drew these concepts together with his proposed cycle of erosion.

Early in his career Chamberlin had been greatly interested in geomorphology, which is the study of landforms, and glaciation. His studies led him to propose that over many millions of years, erosion due to wind, rain, and ice would reduce mountain ranges until they were almost level plains. Rivers would transport the eroded material to the sea, where it would accumulate as sediment. The accumulating sediment would raise the sea level, allowing the sea to advance into low-lying parts of continents. The removal of eroded material would reduce the mass of the continents and they would rise by isostatic readjustment. This vertical movement would raise new mountain ranges that would disrupt the circulation of the atmosphere, producing dry and wet climates in different regions. The changing climate would set in motion the processes of erosion, and the cycle of erosion, isostatic readjustment, mountain formation, and renewed erosion would repeat endlessly.

The idea that the Earth was cooling from an originally high temperature was based partly on the observation that the temperature increases with depth and partly on the prevailing theory of the way planets form. This had been proposed in 1796 in *Exposition du système du monde* (Explanation of the world system), a popular book written by the French astronomer and mathematician Pierre Simon, marquis de Laplace (1749–1827). Laplace believed that a hot cloud of matter, called a nebula, had once rotated about the Sun. This nebula had contracted into rings, which had gradually contracted further to form the planets. Laplace's theory failed to account for the retrograde spin of Venus, however, or the eccentric orbit of the dwarf planet Pluto, or the fact that the Sun emits energy continuously. It was Chamberlin who finally challenged and overthrew the Laplacian theory.

Chamberlin, working in collaboration with the mathematician Forest Ray Moulton (1872–1952), suggested that a star had passed so close to the Sun that their mutual gravitational attraction drew out a stream of matter from one or both bodies. This matter condensed into small, rocky bodies, and as these drew together by gravitational attraction, the planets grew by accretion. Small bodies that orbit a star during the early stages of planet formation are called *planetismals,* and this is known as the planetismal theory of planet formation.

Chamberlin's theory amounted to a direct attack on Lord Kelvin's assertion that the Earth could be no more than 100 million years old because at the end of that time it would have cooled to a frozen, uninhabitable state. Chamberlin criticized Kelvin for sticking dogmatically to a theory that was entirely based on the single hypothesis of a planet cooling from an originally molten state, insisting that scientific theories should grow out of many working hypotheses.

Thomas Chrowder Chamberlin was born in Mattoon, Illinois, on September 25, 1843. His father was a Methodist circuit minister, who augmented his income by farming, as was the custom, and when Thomas Chamberlain was three, the family moved to Beloit, Wisconsin. Chamberlain began his education at a preparatory school before moving to Beloit College, where he learned Latin and Greek, at the same time developing an interest in science. He also directed a church choir. After graduating in 1866, Chamberlin became a teacher at a high school in Beloit and later the school's principal. After two years there, in 1868 he enrolled at the University of Michigan to spend a year studying scientific subjects, including geology, at postgraduate level. From 1869 to 1873 he was professor of natural science at the State Normal School in Whitewater, Wisconsin. He returned to Beloit College in 1873 as professor of geology, zoology, and botany. In 1873 he joined the U.S. Geological Survey, rising to become its chief geologist. While there, between 1877 and 1883 he published *The Geology of Wisconsin.* Chamberlin became president of the University of Wisconsin in 1887. In 1892 he left Wisconsin to head the new geology department at the University of Chicago. He founded the *Journal of Geology* in 1893, and for many years he was its editor. From 1898 to 1914 he was president of the Chicago Academy of Sciences. Chamberlin retired in 1918 and died in Chicago on November 15, 1928.

CONTINENTAL DRIFT

In 1596 a Flemish cartographer, Abraham Ortels (1527–98, also known by the Latinized version of his name, Abraham Ortelius) published an expanded edition of his *Thesaurus geographicus,* a discussion of ancient geography he had first published in 1578 as *Synonymia geographica.* In the 1596 edition Ortels suggests that the Old and New Worlds might once have been joined and had subsequently moved apart. This is the first known suggestion of continental drift. Francis Bacon (1561–1626) also speculated on the topic in his *Novum Organum* (New organon) published in 1620. (*Organon* was the name followers gave to six works on logic by Aristotle.) More recently, in 1858 the French geographer Antonio Snider-Pellegrini (1802–85) proposed in his book *La Création et ses mystères dévoilés* (Creation and its mysteries unveiled) that during the Pennsylvanian epoch (318.1–299 million years ago) all of the continents were joined into a single supercontinent. He based this idea on the fact that fossils of the same plants occurred in Europe and the United States. Others had noted that the continents on either side of the Atlantic looked as though they would fit together.

In 1908 Frank Bursley Taylor (1860–1938), a wealthy American amateur geologist and specialist in the glacial geology of the Great Lakes region, proposed at a meeting of the Geological Society of America that Africa and America had once been joined and that the continents had slowly moved southward from the Arctic. Greenland was left behind when America and Eurasia broke away. Taylor's idea aroused little interest among geologists, but a few years later a German meteorologist and geophysicist, Alfred Lothar Wegener (see sidebar), published a similar idea, this one supported by a vast wealth of evidence. The illustration shows Wegener toward the end of his tragically short life.

Wegener first presented his theory at a Frankfurt meeting of the German Geological Association in 1912. It was published as a paper later that year, and in 1915 it appeared as a book, *Die Entstehung der Kontinente und Ozeane.* Wegener expanded the book for its second

Alfred Lothar Wegener, the German meteorologist and geophysicist who proposed that the continents are moving in relation to each other and that they once occupied positions very different from those they occupy today *(Science Photo Library)*

ALFRED WEGENER: THE GERMAN METEOROLOGIST WHO PROPOSED CONTINENTAL DRIFT

Alfred Lothar Wegener was born in Berlin on November 1, 1880. His father was a minister and director of an orphanage. Alfred Wegener was educated at the Universities of Heidelberg, Innsbruck, and Berlin. In 1905 he received a Ph.D. in planetary astronomy from the University of Berlin but immediately switched to meteorology, taking a job at the Royal Prussian Aeronautical Observatory, near Berlin. He used kites and balloons to study the upper atmosphere and also flew hot air balloons. In 1906, Alfred and his brother Kurt remained airborne for more than 52 hours, breaking the world endurance record.

The same year Wegener joined a two-year Danish expedition to Greenland as the official meteorologist. On his return to Germany, in 1909 he became a lecturer in meteorology and astronomy at the University of Marburg. He collected his lectures into a book published in 1911, *Thermodynamik der Atmosphäre* (Thermodynamics of the atmosphere). This became a standard textbook throughout Germany.

Since 1910 Wegener had been intrigued by the apparent fit of the continental coastlines on either side of the Atlantic Ocean. In 1912 he published a paper, *Die Entstehung der Kontinente und Ozeane* (The origin of the continents and oceans), drawing together various strands of evidence to support the idea he called "continental displacement." This proposed that the continents had once been joined together and have moved slowly to their present positions.

It was also in 1912 that Wegener married Else Köppen, the daughter of Wladimir Peter Köppen (1846–1940), the most eminent climatologist in Germany. Wegener and Köppen collaborated in a book about the history of climate, *The Climates of the Geological Past*. Wegener then returned to Greenland to take part in a four-man 1912–13 expedition. The team crossed the ice cap and was the first to spend the winter on the ice.

At the outbreak of war in 1914, Wegener was drafted into the German infantry but was wounded almost at once. He spent a long time

and third editions. The third edition also appeared in English (as well as in French, Russian, Spanish, and Swedish) as *The Origin of Continents and Oceans* in 1922.

The theory began by recognizing that continents should be seen as rafts of lighter rock floating on denser material. These rafts were able to move and had once been joined in a supercontinent, Pangaea, which began to break apart during the Mesozoic era (251–65.5 million years ago). Wegener pointed out the apparent fit of the coastlines on either side of the Atlantic but explained that this was not simply a matter of their shapes, which might have come to match purely recuperating, during which he elaborated on his theory of continental drift, publishing an expanded version of *Die Entstehung der Kontinente und Ozeane* (*The Origin of Continents and Oceans*) in 1915. The book received a hostile reception from German scientists, and his father-in-law strongly disapproved of Wegener's digression from meteorology into geophysics. Wegener spent the remainder of the war employed in the military meteorological service.

After the war Wegener returned to Marburg. In 1924 he accepted a post created especially for him and became the professor of meteorology and geophysics at the University of Graz, in Austria.

In 1930 he returned to Greenland once more, this time as the leader of a team of 21 scientists and technicians planning to study the climate over the ice cap. They intended to establish three bases, all at 71°N, one on each coast and one in the center, but they were delayed by bad weather. On July 15 a party left to establish the central base, called Eismitte, 250 miles (402 km) inland. The weather then prevented necessary supplies from reaching them, including the radio transmitter and hut in which they were to live. On September 21 Wegener, accompanied by 14 others, set off with 15 sleds to carry supplies to Eismitte. The appalling conditions forced all but Wegener, Fritz Lowe, and Rasmus Villumsen to give up and return. These three finally reached Eismitte on October 30. Lowe was exhausted and badly frostbitten. They stayed long enough to celebrate Wegener's 50th birthday on November 1, then Wegener and Villumsen began their return, leaving Lowe to recover. They never reached the base camp. At first it was assumed they had decided to overwinter at Eismitte, but when they had still not appeared in April, a party went in search of them. They found Wegener's body on May 12, 1931. He appeared to have suffered a heart attack. Villumsen had carefully buried the body. They marked the site with a mausoleum made from ice blocks, later adding a large iron cross. Despite a long search, Villumsen was never found.

There is now an Alfred Wegener Institute for Polar and Marine Research at Bremerhaven, Germany.

by chance. He aligned the geological formations on either side and found that these also matched.

There was still more evidence. Fossils of similar organisms, now extinct, were found in rocks on continents that are now widely separated. Certain other surviving species occur in just a few places separated by vast expanses of ocean they could not possibly have crossed.

Wegener was also interested in *paleoclimatology,* which is the study of ancient climates. He discovered evidence that glaciers had covered parts of Australia, India, Africa, and South America during the Permian period (299–251 million years ago) and maintained that this made sense only if these lands had been grouped around the South Pole during the Permian. Coal measures, made from the remains of plants that once grew in tropical coastal swamps, are found in northern Europe, North America, and other regions far from the Tropics. This must mean, he asserted, that these lands once lay close to the equator.

Wegener described what he believed had happened as "continental displacement" (*die Verschiebung der Kontinente*), and although he built his case on substantial evidence, the evidence was not incontrovertible and his case had weaknesses. He greatly miscalculated the rate at which continents move. From the similarity of glacial moraines in Europe and North America he suggested that the continents had been joined during the Pleistocene ice ages, which ended only about 11,000 years ago. He calculated that Greenland and Europe were moving apart by more than 30 feet (10 m) a year. This was clearly incorrect. Still more seriously, Wegener could think of no convincing mechanism to account for the displacement. He suggested that perhaps tidal forces might have dragged the continents along, or perhaps as the Earth rotated on its axis, inertia made the continents tend to move at a different rate from the solid underlying rocks. These ideas were not convincing, and Wegener knew it.

The geological community was not impressed. Paleontologists preferred the idea that a land bridge had once linked Africa and America, allowing species to cross the ocean, an idea derived from the old theory of a cooling, contracting Earth that Wegener had demolished. Wegener had said the continents began to move during the Mesozoic. This apparently unique event contradicted the generally accepted theory of uniformitarianism—the processes producing change have always operated in the ways they do today—and Wegener could not suggest a reason for it having started. The most powerful criticism related to the magnitude of the forces continental displacement implied. Geologists believed that the layer of rock beneath the continents was extremely dense and did not move; consequently, continents riding like rafts on this material would need to overcome impossibly strong friction. Wegener's ideas were rejected, and for many years they were almost forgotten.

He did have supporters, however. The Canadian-born American geologist Reginald Aldworth Daly (1871–1957), Sturgis Professor of

Geology at Harvard University, thought it feasible that the continents moved by gravity from the poles toward the equator. The most enthusiastic supporter, however, was the South African geologist Alexander Logie du Toit (1878–1948). In 1923 the Carnegie Institution of Washington awarded du Toit a grant, which he used to study the geology of eastern South America. He was greatly struck by the similarities between South American rock formations and those of South Africa, describing them in *A Geological Comparison of South America with South Africa,* published in 1927. In 1937 he published another book, *Our Wandering Continents: An Hypothesis of Continental Drift,* supporting Wegener's theory but changing *continental displacement* to *continental drift,* the term used today. Du Toit did not accept, however, that the continents had moved rapidly, as Wegener had suggested, and he proposed that there had been two supercontinents rather than the single Pangaea. He called these Laurasia and Gondwanaland, the latter the name for the southern continents proposed by Suess (see "Eduard Suess, Colliding Rock Masses, and Moving Continents," pages 174–177). This support was to no avail. Most geologists thought du Toit was simply promoting Wegener's outdated ideas without producing any new evidence to support them. In the end the supporter who did more than any other to rehabilitate Wegener was the English geophysicist Arthur Holmes. Holmes succeeded where others had failed because he discovered how continents move and what drives their movement.

ARTHUR HOLMES—AND THE HOT MANTLE

In the 1890s the controversy over the age of the Earth had still not been resolved. Scientists knew the planet's age had to be reckoned in millions of years, but physicists led by Lord Kelvin favored an age of no more than 100 million years, while geologists believed the Earth to be much older. Holmes was to devote much of his professional life to determining the Earth's true age, and the discoveries he made while doing so led him to propose the mechanism to shift continents. He said that his interest began when he saw in the family Bible that the date of creation had been written as 4004 B.C.E. That puzzled him, because there was no explanation for the precise "4004," rather than a straightforward "4000," and no reason was given for the very young age.

Holmes was born on January 14, 1890, in Low Fell, Gateshead, in the northeast of England. His father was an assistant in a hardware store, and his mother was a schoolteacher. He attended Gateshead higher grade school, where an inspirational teacher, James McIntosh, introduced him to the popular writings of Lord Kelvin and to Suess's *The Face of the Earth,* which had recently been translated into English. When he was 17, Holmes won a scholarship to study physics and mathematics at the Royal College of Science in London (the college became part of Imperial College in 1907 and ceased to exist when it was absorbed into Imperial College in 2002). In his second year he took a course in geology and decided he would become a geologist. He graduated in 1909 in physics, and in 1910 became an associate of the Royal College of Science. He obtained his doctorate in 1917. As well as geology, Holmes loved poetry and history and was a highly accomplished pianist.

Even in 1907 it was not easy to live in London on a scholarship of £60 ($120) a year, and in 1911 Holmes took a job in Mozambique, prospecting for minerals for Memba Minerals, a mining company, but he fell ill with malaria and had to return to England after six months. He obtained work as a demonstrator at Imperial College, remaining there until 1920, when, with his first wife, Margaret, and son, Norman, he went as chief geologist to work for an oil company in Burma (Myanmar). The job lasted only two years. The company failed to pay its employees, and Holmes returned to England bankrupt and devastated by Norman's death from dysentery. Unemployed, he scraped a living giving piano recitals, some say he sold vacuum cleaners, and in partnership with his wife's cousin he opened a shop in Newcastle selling Asian crafts while she sold furs. In 1924 he accepted a post as head of a new geology department at the University of Durham. He was elected a fellow of the Royal Society in 1942, and in 1943 he moved to the University of Edinburgh as Regius Professor of Geology, where he remained until his death in London from pneumonia on September 20, 1965.

In 1904 Rutherford and the American physicist Bertram Borden Boltwood (1870–1927) discovered the steps by which radioactive elements decay to lead and the rate at which this happens (see "How They Built the Geologic Timescale," pages 149–153). This allowed Boltwood to calculate the age of mineral specimens, which he did in 1907, finding they were more than 1 billion years old. Holmes, in collaboration with his lifelong friend Robert Lawson, who was then working at the Vienna Institute of Radium, devised a different method, designed specifically for dating, and found that a specimen of Devonian rock was 370 million years old. Holmes was only 21 when he performed this experiment, and it set the course of his future career. In 1913 he published *The Age of the Earth*, a book in which he argued that the Earth is 1.6 billion years old. The idea was controversial and opposition to it vigorous, but by the early 1920s scientists accepted that the Earth was much older than had once been supposed, and Holmes was recognized as a world authority on the subject.

Holmes was an enthusiastic supporter of Wegener's theory of continental drift. He accepted Wegener's assertion that it explained the rock formations and fossils on either side of the Atlantic and did away with the absurd notion of a land bridge stretching 5,000 miles (8,000 km) between Africa and Brazil to allow animals to migrate. Holmes did more than support the theory, however. His profound understanding of radioactivity and the way rocks form meant he could contribute to it. He knew how much heat radioactivity releases and how slow the process of decay is. He had also studied the way rocks form in the crust. He calculated that differential heating by radioactive decay would set up convection currents within the Earth's mantle, which he called the "substratum." These could generate sufficient force to drag continents, and material from the substratum would rise to fill the gaps this made in the ocean floor. In December 1927 he presented his theory to the Edinburgh Geological Society in a paper called "Radioactivity and Geology."

Opposition to his idea was widespread and vociferous, both in Britain and the United States. Nevertheless, Daly, professor of geology at Harvard (see "Continental Drift," pages 185–189), invited Holmes to the United States in 1932 to deliver the Lowell lectures on geology and radioactivity. It was not enough to win acceptance from the geological community. That did not come until the 1960s, when geologists realized that his theory of mantle convection had been generally correct.

In 1944 Holmes was commissioned by the British government to write a book on physical geology for Royal Air Force officer cadets. He wrote *Principles of Physical Geology* while firewatching, which was a task most noncombatant city dwellers were required to perform during World War II: Standing in a high position, often the roof of a tall building, during air raids, they telephoned the locations of fires to the fire brigade. His book became a classic, always known simply as *Holmes.* He rewrote it entirely in 1965, and the 1965 edition was reprinted six times. Holmes's second wife, Doris Reynolds, wrote a third edition published in 1978, and her student P. Donald Duff wrote a fourth edition published in 1993.

PALEOMAGNETISM

In 1849 Achilles Delasses, a French physicist, discovered that certain rocks were magnetized, their magnetic fields being aligned with the Earth's magnetic field (known as the geomagnetic field). In 1906 another French physicist, Bernard Brunhes (1867–1910), found other rocks that were magnetized but with fields aligned in the opposite direction to the geomagnetic field. This led Brunhes to suggest that the geomagnetic field had reversed its polarity at some time in the past. Magnetic north had become south, and south had become north. These scientists had discovered *ferromagnetism* and also evidence that the geomagnetic field can change its polarity.

Iron, nickel, and cobalt possess *ferromagnetic* properties, which means they can be magnetized to form permanent magnets. Alloys or minerals containing any of these metals can also be magnetized. Rocks containing such minerals consequently behave like very weak but permanent magnets. While rocks are molten, atoms of ferromagnetic metals inside them are free to orient themselves in any direction, but as the temperature falls they align themselves with the geomagnetic field and below a certain temperature, known as the Curie temperature, the atoms remain locked in that alignment. They can no longer be magnetized by a magnetic field. The rocks then record the alignment of the geomagnetic field as it was when they cooled below their Curie temperature. For hematite the Curie temperature is 1,247°F (675°C) and for magnetite it is 1,067°F (575°C). The magnetism the rocks retain is called *natural remanent magnetization.*

Mariners had been using magnetic compasses since the 11th century in China and since about 1300 in Europe, so scientists were very familiar with the Earth's magnetic field. They knew that the field is approximately aligned with the planet's rotational axis but that its alignment and strength vary. The geomagnetic poles are some distance from the geographic poles. But no one knew how the geomagnetic field is generated. This was debated among physicists and geophysicists in the 1950s. Most scientists believed (and still do) that convection currents in the Earth's partly liquid, iron-rich core generate the magnetic field. Others held that any rotating body will generate a magnetic field. That was the view of the English astrophysicist Patrick M. S. Blackett (1897–1974), who won the 1948 Nobel Prize in physics. Blackett's view was eventually rejected, but it raised a further point. If Blackett was correct, then the location of the magnetic poles should not change, and the natural remanent magnetization in rocks should record the direction of the geomagnetic field at the time they cooled below their Curie temperature. When the evidence of paleomagnetism in the rocks was assembled, however, it showed that if Blackett was right and the geomagnetic field had not moved, then during the last 200 million years Britain had moved through 30 degrees of longitude and a considerable distance northward.

The English geophysicist S. Keith Runcorn (1922–95) interpreted the data differently. He calculated former positions for the geomagnetic poles, as measured from Britain, and plotted them as a polar wandering curve, which he published in 1955, saying that this showed it was the poles that had moved and not the land. When a similar curve was plotted for North America, it was similar to the British one, with one difference. The North American alignments placed the magnetic North Pole to the West of the position indicated by the British data. This could only mean that North America and Europe had once been closer together than they are at present. It strongly suggested that both the poles and the continents had moved.

In the 1960s studies of natural remanent magnetization were conducted on samples of rocks from many parts of the world. When the results were compared, it became clear that the geomagnetic field had reversed its polarity nine times in the last 3 million years. More recent studies probing further into the past have found more than 170 reversals in the last 76 million years. Geophysicists have now constructed a global polarity timescale, dividing Earth history into intervals called polarity superchrons, chrons, and subchrons.

These vary greatly in length. There were no reversals, for example, during the Cretaceous superchron, lasting from 120 to 83 million years ago, or during the Kiaman reversed superchron, from about

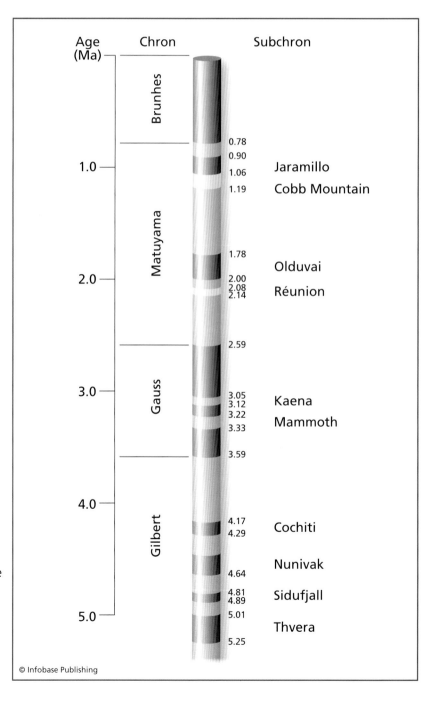

Geomagnetic polarity reversals. The diagram shows the polarity reversals of the Earth's magnetic field over the last few million years. The dark areas indicate periods of normal polarity (polarity as it is at present, with the geomagnetic north pole in Canada), and the light areas indicate reversed polarity.

© Infobase Publishing

316 to 262 million years ago, when polarity was reversed. The diagram shows the polarity chrons and polarity subchrons during the last few million years.

ROBERT DIETZ—AND SEAFLOOR SPREADING

Jaramillo Creek, in New Mexico, is the source of the rocks that in 1965 a team of scientists from the University of California at Berkeley, led by Richard Doell (1923–2008), G. Brent Dalrymple (1937–), and Alan Cox (1926–1987), used to identify the most recent polarity reversal. Their discovery completed a picture of what had been happening over millions of years on the ocean floor. It was the final piece of evidence that proved the reality of seafloor spreading.

Ocean research had intensified during World War II. Navies searching for enemy submarines and seeking to conceal their own needed to map the ocean floor and investigate its topography, and they stimulated the rapid development of the technologies to achieve this. Knowledge of the natural remanent magnetization in the rocks of the ocean floor and the ability to detect and measure it made submarine detection easier, and the U.S. Navy devoted considerable effort to oceanographic research. The impetus to research that began as a response to military needs continued after the war ended. Many of the scientists involved in that research effort had served at sea, but it was a former pilot in the U.S. Army Air Corps, Robert S. Dietz (1914–95), who coined the term *seafloor spreading* in 1961.

The wartime studies brought many surprises. Scientists discovered that vast areas of the ocean floor are a level and featureless abyssal plain covered by sediment lying on basalt rocks averaging only three miles (5 km) thick. If the Earth's surface was static, as all but a few geologists believed, then the rocks of the ocean floor should be very ancient, but they were not. Scientists discovered that the uniformity of the abyssal plain is broken by patterns of deep trenches and submarine mountains forming ridges. These mountain chains rise to an average height of 14,765 feet (4,500 m) above the ocean floor and in places are more than 500 miles (800 km) wide. The mid-ocean ridges found in all the oceans and virtually encircling the planet are made of recently cooled lava. These are the youngest rocks. To either side of the ridges the natural remanent magnetization formed a pattern of bands, parallel to the ridges, of alternating normal and reversed polarity. The Jaramillo subchron was the band closest to the ridges. The picture that finally emerged showed that the rocks forming the ocean floor were much younger than continental rocks, and the youngest rocks of all were those closest to the mid-ocean ridges.

Dietz was born in Westfield, New Jersey, on September 14, 1914. From 1933 to 1941 he studied geology and chemistry at the University of Illinois. He conducted most of his doctoral research in marine geology at the Scripps Institution of Oceanography in California. After obtaining a Ph.D. from Illinois he was drafted into the U.S. Army Air Corps, and he remained in the reserves for 15 years after the war ended, retiring with the rank of lieutenant colonel.

After the war Dietz founded and became the director of the Sea Floor Studies Section of the Naval Electronics Laboratory, in San Diego. In 1955–56 he sailed as geological oceanographer on Operation Highjump, the last expedition to Antarctica led by Admiral Richard E. Byrd (1888–1957), and he participated in several studies of the Pacific Basin. His studies of the ocean floor and in particular of the volcanic activity along the mid-ocean ridges led Dietz to realize how ocean basins form. He called the process "seafloor spreading," in which the seafloor on either side of a mid-ocean ridge moves away from the ridge and the gap between the sections of seafloor is filled by rock rising from the mantle. Dietz described it as a conveyor belt driven by convection currents in the mantle that piles up rocks rich in silica and aluminum. These piled rocks become continents that are carried away from the mid-ocean ridges, and the ocean floor and upper mantle make up a thin layer, the *lithosphere,* above the *asthenosphere,* the slightly plastic layer on which the continents ride.

Dietz made one other point: Continents result from compression as rocks are piled together, and ocean basins result from tension as the seafloor expands, but the Earth as a whole neither contracts nor expands. This was significant, because many geologists supposed that continental drift could be explained if the Earth had expanded greatly in size since it formed. The discovery that new mantle material is constantly being added to the ocean floor fits well with this theory. The difficulty was that no one could suggest any credible mechanism that might bring about such an expansion. Dietz and the geologist Harry Hammond Hess rejected the idea that the Earth changed its size. They realized that if new material is being added to the crust in one place, material must be leaving the crust somewhere else—in the ocean trenches.

Dietz died of a heart attack at his home in Tempe, Arizona, on May 19, 1995.

HARRY HESS—AND MID-OCEAN RIDGES

The ocean floor expands from the mid-ocean ridges, and Hess (1906–69) was the geologist who first explained how this happens. In doing so Hess answered a question that had been puzzling many geologists for some time. Scientists knew that the oceans have existed for at least 4 billion years. Presumably, therefore, material washed from the continents has been accumulating as sediment on the ocean floor for the whole of that time. Why is it, then, that the layer of sediment covering the ocean floor is relatively thin? Hess calculated that sediment has been accumulating only for as long as it takes for the ocean floor to expand from the mid-ocean ridges to the trenches; consequently, nowhere was the ocean floor more than 300 million years old, and that is how long the sediment had been accumulating. In other words, the ocean floor is constantly recycling. Recycling also explains why no fossils older than about 180 million years have ever been found in seafloor sediments, although fossils of marine organisms many millions of years older than that occur in sedimentary rocks that have been thrust upward to form mountains, for example, in the Alps and Himalayas.

In 1962 Hess described how seafloor spreading works in a book entitled *History of Ocean Basins,* based on an expanded version of a report he made in 1960 to the Office of Naval Research. Hess proposed that new ocean floor emerges at mid-ocean ridges, and ocean floor disappears at the edges of the oceans where it is subducted into the mantle. The recycling of ocean floor provides the mechanism for the movement of continents. The continents are carried passively by the expansion of the ocean floor, rather than having to plow through the dense rock of the static oceanic crust, as Wegener had suggested.

Hess was born in New York City on May 24, 1906. He attended Asbury Park High School in New Jersey and entered Yale University in 1923 to study electrical engineering but after two years changed to geology, graduating in 1927. Hess then spent two years working as an exploration geologist in Northern Rhodesia (now Zambia) before returning to Princeton University, where he obtained his doctorate in 1932. He taught at Rutgers University for one year, moved to the Geophysical Laboratory of the Carnegie Institute of Washington as a research associate for a year, and in 1934 he returned to Princeton where he spent most of his career. From 1950 to 1966 he was chair of the university's geology department, and from 1964 he was Blair Professor of Geology. He was also a visiting professor at the University of Cape Town, South Africa, in 1949–50 and at Cambridge University in 1965.

During the 1930s Hess took part in submarine gravity surveys around the West Indies under the Dutch geophysicist Felix Andries Vening Meinesz (1887–1966), who had perfected the technique for making accurate gravity measurements in the stable environment of a submarine, sailing in submarines of the Royal Netherlands Navy. Hess later extended these studies to the Lesser Antilles and to facilitate access to work on board submarines of the U.S. Navy he joined the naval reserve as a lieutenant second grade. He eventually attained the rank of rear admiral.

In 1941 Hess was called to active service and devised a system for estimating the positions of German submarines in the North Atlantic, testing his method on the submarine decoy vessel USS *Big Horn.* He later commanded the attack transport USS *Cape Johnson* in the Pacific, where he took part in major landings on the Marianas, Philippines, and Iwo Jima. With the agreement of his crew, on the way across the Pacific to these landings Hess carefully chose a route that would allow the *Cape Johnson*'s echo-sounding equipment to study the ocean floor and in particular the flat-topped volcanic seamounts he named *guyots* in honor of the Swiss-American geographer Arnold Guyot (1807–84), who had founded the Princeton geology department.

In later years Hess received many honors. He was elected to the National Academy of Sciences, the American Philosophical Society, the American Academy of Arts and Sciences, and the geological societies of London, South Africa, and Venezuela. He served as president of the Mineralogical Society of America and the Geological Society of America. The American Geophysical Union established the Harry H. Hess Medal, awarded for "outstanding achievements in research in the constitution and evolution of Earth and sister planets."

In 1962 President John F. Kennedy appointed Hess as chair of the Space Science Board of the National Academy of Sciences. Hess died of a heart attack on August 25, 1969, while presiding over a conference of the Space Science Board he had organized at Woods Hole, Massachusetts, to revise the scientific objectives of lunar exploration. He was buried at Arlington National Cemetery and was posthumously awarded the National Aeronautics and Space Administration's Distinguished Public Service Award.

FRED VINE, DRUMMOND MATTHEWS, AND PLATE TECTONICS

Hess knew that it was seafloor spreading that made continents move, but there was insufficient evidence to prove it. He explained the process in a paper he published in 1960, but because the evidence was weak, he called it "An Essay in Geopoetry." The evidence appeared during the 1960s with the introduction of new devices for studying the deep ocean floor.

In 1964 the Canadian geophysicist John Tuzo Wilson (1908–93) took part in a symposium on continental drift sponsored by the Royal Society and held in London. Wilson had once believed in the cooling and contracting theory of the Earth, but the discovery of paleomagnetism and Hess's "geopoetry" had in combination persuaded him of the truth of seafloor spreading. He saw further evidence for it at the symposium when Sir Edward Bullard (1907–80), professor of geodesy and geophysics at Cambridge University, presented a computer-generated map showing how closely the coasts of Africa and South America fitted together.

Edward Crisp Bullard and the American William Maurice Ewing (1906–74) are credited with having founded the scientific discipline of geophysics. Bullard was knighted in 1953, and in 1968 he received the Vetlesen Medal and Prize, equivalent to a Nobel Prize in Earth sciences. Bullard's reputation attracted talented young scientists to Cambridge, among them Drummond Hoyle Matthews (1931–97), a research fellow, and his graduate student Frederick John Vine (1939–). As part of the 1961–63 International Indian Ocean Expedition, in 1962 Matthews and Vine conducted an ocean survey in the Gulf of Aden along part of the mid-ocean ridge of the Indian Ocean. During cruises on the HMS *Owen* they observed patterns of magnetic stripes, up to 20 miles (32 km) wide, similar to those found earlier in the Pacific. The stripes, running symmetrically on either side of the ridge, recorded reversals in geomagnetic polarity, and Matthews and Vine realized that the stripes also recorded the spread of the ocean floor from the central ridge. Their paper describing their findings, called "Magnetic Anomalies over Ocean Ridges," was published in 1963 in *Nature*.

Early in 1965 Wilson, Hess, Matthews, and Vine met at Cambridge to discuss seafloor spreading and its implications. In 1966 Matthews took over as head of the Cambridge marine geophysics group. Between 1967 and 1982 the group's 15 members took part in more than 70 seagoing expeditions. Vine became dean of the School of Environmental Science at the University of East Anglia. He retired in 1998 and became Professor Emeritus. Wilson went on to discover hotspots and showed that the Hawaiian Islands are on a tectonic plate that is drifting in a northwesterly direction over a fixed hotspot in the ocean floor. The hotspot produced volcanoes forming a long chain of volcanic islands on the oceanic crust. Wilson also discovered transform faults, which he interpreted as fractures in the rocks on either side of a ridge as the crust was torn apart.

At first the Matthews and Vine paper aroused little interest. Geomagnetic polarity reversal had not yet been proved, and many scientists thought it improbable. When, in 1963, the Canadian geophysicist Lawrence W. Morley (1920–   ) independently proposed the theory of magnetic imprinting on the ocean floor, his paper was rejected with the comment that although such ideas might make for interesting conversation at cocktail parties, they were not suitable for publication in a serious scientific journal. Morley later directed the Canada Centre for Remote Sensing and became Canadian Science Counsellor at the Canadian High Commission in London.

The American geophysicist Lynn R. Sykes (1937–   ) of the Lamont-Doherty Earth Observatory at Columbia University is another winner of the Vetlesen Medal and Prize. Wilson had suggested that seafloor spreading could explain strange seismic records from certain parts of the East Pacific Rise. Sykes studied the detailed records of 20 earthquakes from different parts of the world and confirmed Wilson's idea. In every case the rocks moved in the opposite direction to that which geologists would ordinarily expect. He presented his results to the Geological Society of America in 1966. After that more evidence began to arrive rapidly. Finally, in 1968, the age of sediments along 30-foot (9-m) cores of rock drilled from the floor of the Atlantic was measured. This revealed that the age of the sediment lying immediately above the basalt basement rock was directly related to its distance from the ridge. It proved that seafloor spreading is happening and its rate. In the Atlantic this is 0.75 inch (1.9 cm) a year on either side of the ridge. It proved that the Atlantic Ocean is growing wider by 1.5 inches (3.8 cm) a year.

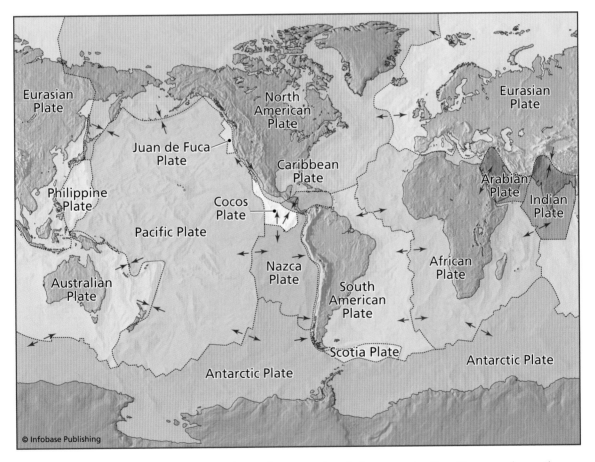

Earth's crust is broken into sections, or plates, that move in relation to each other. The map shows the seven major plates (African, Antarctic, Eurasian, Indian, North American, Pacific, and South American) and several of the minor plates. The arrows indicate whether the plates are diverging (divergent boundary), converging (convergent boundary), or moving past each other in opposite directions (conservative boundary).

As all of these ideas came together, the modern picture of the Earth's crust emerged as the unifying theory called plate tectonics (*tectonism* means "deformation"). According to the theory of plate tectonics, the crust is broken into sections, called "plates," which move in relation to each other, carried by convection currents in the mantle. Oceans grow wider, spreading from mid-ocean ridges, or narrower as dense oceanic crust sinks into the mantle beneath lighter continental crust at boundaries where plates collide. Subduction, which is the process in which one plate is drawn downward beneath another, is what drags sections of crust along. Continents also collide with other continents. Plates move at different rates.

Geophysicists recognize seven major tectonic plates: the African, Antarctic, Eurasian, Indian, North American, Pacific, and South American plates. There are also minor plates such as the Filipino, Juan de Fuca, Nazca, Cocos, Caribbean, Arabian, and Scotia plates, as well as 14 microplates, including the Carline, Burma, Tonga, and Galápagos microplates. The plates range in size from the Pacific plate, with an area of 39,874,000 square miles (103,300,000 km^2), to the Galápagos microplate, with an area of 4,632 square miles (12,000 km^2). The map shows the major and minor plates.

In some places plates have fused together along sutures marked by highly deformed rocks and mountain building. Active boundaries between plates, where the plates are still moving, can be divergent, convergent, or conservative. Divergent boundaries are where plates are moving apart at a mid-ocean ridge. Convergent boundaries are where plates are colliding—oceanic crust with oceanic crust, oceanic crust with continental crust, or continental crust with continental crust. At conservative boundaries plates are moving past each other in opposite directions. The arrows in the illustration identify divergent, convergent, and conservative margins.

Plate tectonics is a truly unifying theory, because every aspect of the geologic sciences relates to it. It explains why earthquakes and volcanism occur in certain places—along plate boundaries—how the mountain ranges are formed, why there are sedimentary rocks containing marine fossils high in the Alps and Himalayas, and why the Highlands of Scotland are geologically so similar to parts of North America.

Conclusion

Today satellites orbit our planet. Their images map not only the coastlines of the continents and islands, the mountains and plains, lakes and rivers but also the changing appearance of landscapes. They mark the advance and retreat of winter ice, the clearing of forests, the droughts that wither crops, and the rains that green the land. They chart the formation and drift of weather systems, revealed by cloud patterns, and quietly monitor dust storms and tropical cyclones, allowing meteorologists to warn those in their path. The satellite-based Global Positioning System tells the traveler of his or her position, however remote. We have truly mastered the geography of Earth.

But below the surface the quest for knowledge continues. Geologists search for new sources of fuels and minerals. Seismologists study the mechanisms of earthquakes in the hope that one day they will be able to predict their occurrence in time for people to escape. Volcanologists seek the understanding of volcanism that will allow them to provide reliable advance warning of eruptions.

Plate tectonics is the theory that underlies all of the modern Earth sciences, but although its general principles are understood, geophysicists still seek to fill in the details of plate movements. Plate movements have widespread consequences. The Asian monsoon, on which harvests depend, is caused partly by the flow of air over and between the Himalayan mountain ranges, which have been thrust upward by the collision between the Indian and Eurasian plates.

There are entire libraries of books that describe particular branches of the Earth sciences in varying detail and at different levels of complexity. A single book as short as this one can do no more than provide the briefest introduction, a quick excursion around parts of the overall topic. This book has focused on the discoveries that have been made and how knowledge has increased over many centuries.

Earth sciences have developed because humans are by nature inquisitive, driven to find explanations for the phenomena they observe. Over the centuries this quest for understanding has given rise to a method of study loosely known as the scientific method. This is not in the least mysterious. It simply consists of persistent, patient, careful observation and rigorous techniques for testing possible explanations of what has been observed. The study of the Earth will continue for as long as there are questions to be answered and scientists to ask them.

GLOSSARY

abiogenesis (spontaneous generation) the theory that living organisms can emerge directly from nonliving matter, for example, that dirty wheat grains give rise to mice.

alloy a mixture of two or more metals.

alluvial pertaining to a river.

annealing making **wrought iron** malleable by heating it and allowing it to cool slowly.

asthenosphere the weak zone in the upper **mantle,** immediately beneath the **lithosphere.**

atomic number *see* **nuclide.**

blister steel steel produced by the **cementation** process.

bloomery the earliest type of iron smelter.

body wave a **seismic wave,** either a P-wave or S-wave.

bronze an **alloy** of copper and tin.

cartography map drawing.

cast iron iron with a carbon content of 2–6 percent that has been melted and shaped by casting it in a mold. After cleaning it can be shaped slightly by hammering, but it is very brittle.

catastrophism the theory that the Earth's history has been punctuated by violent (catastrophic) events producing mass extinctions.

cementation a process for producing blister steel by heating a clay box containing bars of **wrought iron** packed with charcoal for several days.

clepsydra a clock that works by allowing water to drip into a container at a controlled rate, raising a float attached to which there is a pointer marking a position on a drum.

coppicing the woodland management technique of cutting broad-leaved trees close to ground level in order to encourage the growth of thin poles from around the edge of the stump.

Coriolis effect (CorF) the apparent deflection, to the right in the Northern Hemisphere and to the left in the Southern Hemisphere, experienced by any body moving across the Earth's surface but not attached to the surface. It is due to the rotation of the Earth beneath the moving body. Its magnitude varies with latitude and the speed of the moving body; it is zero at the equator and at its maximum over the poles.

Damascus steel   a high-quality steel, based on **wootz steel,** that was made in the Middle East from about 1100 to 1700.

dead reckoning   a navigational method for estimating position based on measuring the time that has elapsed since the last observational determination of position and the speed and direction of travel.

decay   the spontaneous transformation of a radioactive **nuclide** into a daughter nuclide with the emission of one or more particles and/or radiation.

decay constant   the rate at which a radioactive element emits radiation and decays.

declination   the angle of a celestial object above the horizon, measured in degrees, minutes, and seconds.

degrees of freedom   the number of independent parameters that are needed to describe the way a system is configured. For example, the atoms and molecules of a substance move in several different ways; these are the degrees of freedom that must be considered when calculating the way the substance may respond to absorbing heat energy.

diluvialism   the theory that on one or more occasions the entire Earth has been flooded.

disjunct distribution   the situation where closely related organisms occur naturally only in a few places separated by an insuperable barrier such as an ocean or mountain range.

electropositive   carrying a positive electromagnetic charge.

ellipsoid   *see* **spheroid.**

epicenter   the place on the Earth's surface that is directly above an earthquake **hypocenter.**

equinox   one of the two days each year (presently March 20–21 and September 22–23) when the noonday Sun is directly overhead at the equator, and dawn and sunset are 12 hours apart everywhere in the world.

erratic   a bed of gravel and boulders transported by a **glacier** far from their original location.

eustasy   a worldwide change in sea level caused either by tectonic movements or by the growth or melting of **glaciers,** called glacioeustasy or glacioeustasism.

ferromagnetic   possessing the property of **ferromagnetism.**

ferromagnetism the magnetism found in iron, nickel, cobalt, and certain other metals and in **alloys** and **minerals** containing them.

flux a substance added to a smelter to make the molten ore more fluid and to facilitate the separation of the required metal.

focus *see* **hypocenter.**

geochemistry the scientific study of the abundance and distribution of the chemical elements in the Earth and solar system, and their natural circulation on Earth through geochemical cycles.

geodesy the scientific study of methods of surveying and mapping the Earth's surface, measuring the Earth's size and shape, and measuring its gravitational field.

geoid the shape of the Earth given by a surface that is everywhere perpendicular to the direction of gravity; it corresponds to mean sea level if the sea were spread across the entire Earth, including the continents.

geomorphology the study of the landforms of the Earth and the process that forms them.

geosyncline a large depression in the Earth's crust caused by the downward movement of crustal rocks. Prior to the introduction of the plate tectonics theory this was considered a major cause of mountain formation and other geologic features. Today the term describes a linear trough filled with sediment that forms sedimentary rock and is subsequently uplifted to form a belt of folded mountains.

glacier a large mass of ice lying on or adjacent to the land surface and usually moving. Valley glaciers flow down mountain sides; ice sheets cover areas of more than 9,650 square miles (25,000 km^2); ice caps are similar to ice sheets but smaller.

glacioeustasism *see* **eustasy.**

glacioeustasy *see* **eustasy.**

great circle the shorter arc of a circle drawn on the surface of a sphere whose center is coincident with the center of the sphere.

groundwater water that has drained from the ground surface and accumulated above a subterranean layer of impermeable material.

half-life the time taken for half of the atoms of a radioactive element to **decay.**

humidity the amount of water vapor present in the air.

hygrometer an instrument for measuring atmospheric **humidity.**

hypocenter (focus) the place in the Earth's crust where the rock movement responsible for an earthquake happens.

ice cap   *see* **glacier.**

ice sheet   *see* **glacier.**

igneous   of a rock formed from the solidification of molten rock.

inclination   *see* **magnetic dip.**

inverse square law   the physical law, discovered by Isaac Newton (1642–1727), stating that the force of gravitational attraction between two bodies is inversely proportional to the square of the distance between them.

ion   an atom or group of atoms that has lost or gained one or more electrons.

isotope   a variety of an element in which the atoms contain the same number of **protons** as all other atoms of that element but a different number of **neutrons.** This alters the mass of the atom but not its chemical properties.

law of faunal succession   the principle, discovered by William Smith (1769–1839), that rock strata over large areas can be identified by the assembly of fossils that each stratum contains and that the fossil assemblages succeed one another in a definite, regular order.

law of original horizontality   the principle that **sedimentary rocks** formed from sediments deposited in horizontal or near-horizontal layers (strata).

law of superposition of strata   the principle that in a **sedimentary rock** the stratum that was originally at the bottom of the sequence is the oldest, those above it are progressively younger, and the youngest stratum is the one at the top. These positions may be changed by subsequent folding of the rocks, so geologists studying a stratigraphic sequence must determine the original orientation (the "way up") of the strata.

lithosphere   the Earth's crust and the uppermost, brittle part of the **mantle.**

magnetic dip (inclination)   the angle a suspended magnet makes with the horizontal if it is allowed to move vertically. This angle is 0° at the equator, +90° at the North Pole, and –90° at the South Pole.

mantle   the region of the Earth lying between the outer core and the base of the crust.

martensite   a solid solution of carbon in iron.

mean time (mean solar time)   the time measured with reference to a notional Sun that crosses the sky in a circular orbit at a constant speed, with a day that is precisely 24 hours long; time zones are based on mean time.

meridian   a line from the center of the horizon in the direction of true (not magnetic or compass) north, passing directly over the head of the observer and ending at the horizon to the rear.

meridian arc   a section of the **meridian.**

mineral   a naturally occurring, usually inorganic substance, with a characteristic chemical composition and typically a crystalline structure; rocks are made from minerals.

natural remanent magnetization   the magnetism possessed by certain rocks, acquired by natural processes.

Neptunism   the theory that rocks of the Earth's crust formed beneath an ocean that once covered the entire planet, although not all of the surface was necessarily covered at the same time.

neutron   an atomic particle carrying no net charge.

nuclide   an atomic nucleus, characterized by the number of **protons** it possesses (its atomic number) and the number of **neutrons.**

ore   a rock or **mineral** that contains a metal that can be extracted commercially.

orogeny   an episode of mountain building.

paleoclimatology   the study of ancient climates.

paleontology   the study of fossil plants and animals.

permafrost   a layer of ground that remains permanently frozen. For a permafrost layer to develop the ground must remain frozen for at least two winters and the intervening summer.

phase   a distinct, homogeneous form within a system consisting of two or more such forms; a mixture of ice and liquid water constitutes a two-phase system.

pig iron   iron alloyed with carbon and other impurities.

planetismal   by international agreement, a solid object arising during the accumulation of planets whose internal strength is dominated by its own gravity and whose orbital dynamics are not significantly affected by drag from surrounding gas. This corresponds to objects larger than approximately 0.6 mile (1 km) in the solar nebula.

Plutonism   the theory that subterranean fires cause volcanism, which produces new rock that then erodes away.

precession of the equinoxes   the rate at which the Earth's position in its solar orbit at the equinoxes changes over time; the rate of precession is 50.27" per year.

projection   in **cartography,** the technique of transferring features on a spherical surface onto a two-dimensional surface by identifying control points and projecting their locations onto the two-dimensional sheet, then drawing the features in relation to them.

proton   an atomic particle carrying positive charge.

radiometric dating   the use of the known **decay constant** of a radioactive **nuclide** to determine the age of a sample of rock or other material.

relative humidity   the amount of water present in the air expressed as the percentage of the amount needed to saturate the air at that temperature.

rhumb line   a line drawn on a map that crosses all **meridians** at the same angle.

rock burst   an explosion of the hard rock lining the sides and roof of galleries in deep mines caused by changes in pressure due to the removal of rock combined with cooling to lower temperatures to levels tolerable for miners.

sedimentary rock   a rock formed from sediments that have been hardened by compression and often heating.

seismic wave   a shock wave that travels through the Earth from the site of an earthquake or explosion.

seismograph   a device that records **seismic waves** electronically and as an ink trace on a rotating drum.

seismology   the scientific study of earthquakes and the **seismic waves** they produce.

seismometer   a device that detects **seismic waves.**

sidereal time   the time measured with reference to the fixed stars; a sidereal day is 23 hours 56 minutes 4.09 seconds long, and there are 366.24 sidereal days in a sidereal year.

slag   the substance produced by a chemical reaction during **smelting** between a **flux** and impurities in an **ore.** The slag floats on top and can be separated by pouring it away.

smelting   separating a metal from its **ore** by heating the ore in a furnace.

solstice   one of the two days each year (presently June 21–22 and December 22–23) when the noonday Sun is directly overhead at the tropic of Cancer (June) or Capricorn (December); it is Midsummer Day in the illuminated hemisphere and Midwinter Day in the opposite hemisphere.

sonar   an echo-sounding technique for locating objects and topographic features on the seafloor by emitting a sound wave and measuring the time that elapses between emission and receipt of the echo. The name is short for *sound navigation ranging*.

spheroid (ellipsoid)   the three-dimensional figure that is formed by rotating an ellipse about its longer (semimajor) axis.

spiegel (spiegeleisen)   a form of **pig iron** containing 15–30 percent manganese and 4–5 percent carbon, that is added to the molten metal during the Bessemer process.

spontaneous generation   *see* **abiogenesis.**

stratigraphic column   a succession of **sedimentary rock** strata laid down during a specified period.

stratigraphy   the branch of Earth science concerned with the study of stratified rocks.

tsunami   a shock wave caused by a submarine earthquake, volcanic eruption, or sediment slide that travels at high speed across the ocean, its height increasing as the wave enters shallow water and slows.

unconformity   a surface where two rock strata, formed at different times, make contact.

water table   the upper surface of the **groundwater.**

wootz steel   steel made in India and Sri Lanka by a crucible process from about 300 B.C.E.

wrought iron   iron with a low carbon content (about 0.04 percent) that is shaped by hammering, stretching, and twisting either hot or cold.

zenith   the point directly overhead.

FURTHER RESOURCES

BOOKS AND ARTICLES

Allaby, Michael, and Derek Gjertsen. *Makers of Science.* Vols. 1–3. New York: Oxford University Press, 2002. Books in this five-volume set consist of brief biographies of leading scientists and summaries of their contributions. Volume 1 has information about Aristotle, Galileo Galilei, and Antoine Lavoisier. Volume 2 has information about Alexander von Humboldt. Volume 3 has information about Alfred Wegener.

Allaby, Michael, and James Lovelock. *The Great Extinction.* Garden City, N.Y.: Doubleday, 1983. An account of the unraveling of evidence of a major impact event at the Cretaceous-Tertiary boundary that caused a mass extinction.

Bottke, William F., David Vokrouhlicky, and David Nesvorny. "An asteroid breakup 160 Myr ago as the probable source of the K/T impactor." *Nature* 449 (2007): 48–53. A paper describing the likely origin of the body responsible for the Cretaceous-Tertiary impact.

Bowler, Peter J. *The Fontana History of the Environmental Sciences.* London: HarperCollins, 1992. A general history of the environmental sciences by a leading historian.

Danson, Edwin. *Weighing the World: The Quest to Measure the Earth.* New York: Oxford University Press, 2006. The story of how the mass of the Earth was measured and how a Scottish mountain was weighed.

Feresin, Emiliano. "Fleece myth hints at golden age for Georgia." *Nature* 448 (August 23, 2007): 846–847. Paper suggesting a possible origin for the legend of the Golden Fleece.

Hancock, Paul L., and Brian J. Skinner, ed. *The Oxford Companion to the Earth.* New York: Oxford University Press, 2000. A comprehensive guide to Earth sciences.

Holdeer, Gerald D., and P. R. Bishnoi, eds. *Gas Hydrates: Challenges for the Future* Vol. 912. New York: Annals of the New York Academy of Sciences, 2000. Report of a conference on methane hydrates, with contributions from many authors.

Jardine, Lisa. *Ingenious Pursuits: Building the Scientific Revolution.* London: Little, Brown, 1999. A history of the development of science in Europe in the 17th and 18th centuries by a leading historian.

Kumar, Prakash, et al. "The rapid drift of the Indian tectonic plate." *Nature* 449 (October 18, 2007): 894–897. A paper on the movement of the Indian plate.

Longwell, Chester R. *Clarence Edward Dutton, 1841–1912, a Biographical Memoir.* Washington, D.C.: National Academy of Sciences, 1958. A brief biography of Dutton.

Polo, Marco. *The Travels.* Trans. Ronald Latham. Harmondsworth, U.K.: Penguin Books, 1958. The famous account of the travels and adventures of Marco Polo, originally written in 1298–99.

Rudwick, Martin J. S. *Bursting the Limits of Time: The Reconstruction of Geohistory in the Age of Revolution.* Chicago: University of Chicago Press, 2005. A major work describing the way the age and history of the Earth were discovered.

Russell, Bertrand. *History of Western Philosophy.* New York: Taylor & Francis, 2004. A classic work on the history of European thought.

Schaefer, Bradley E. "The Origin of the Greek Constellations." *Scientific American* 295, no. 5 (November 2006): 70–75. An article describing how the Greeks named the constellations.

WEB PAGES AND SITES

Aber, James S. "Birth of the Glacial Theory." In *History of Geology.* Emporia State University. Available online. URL: http://academic.emporia. edu/aberjame/histgeol/agassiz/glacial.htm. Accessed September 13, 2007. An account of the development of the theory of ice ages.

———. "Georgius Agricola." In *History of Geology.* Emporia State University. Available online. URL: http://academic.emporia.edu/aberjame/ histgeol/agricola/agricola.htm. Accessed June 21, 2007. A brief biography of Agricola.

About Maps. "The Early History of World Maps." Moonville Studios. Available online. URL: http://www.maps-gps-info.com/history-of-world-maps.html. Accessed March 5, 2007. A description of some of the world's earliest surviving maps.

Davis, Henry. "Orbis Terrarum." Ancient Web Pages. Available online. URL: http://www.henry-davis.com/MAPS/AncientWebPages/118mono. html. Accessed February 5, 2007. A description of the map of the Roman Empire made by Marcus Vipsanius Agrippa.

Enersen, Ole Daniel. "Johann Friedrich Blumenbach." Who Named It? Available online. URL: http://www.whonamedit.com/doctor. cfm/1247.html. Accessed September 19, 2007. A brief biography of Blumenbach.

Glasby, Geoff. "Goldschmidt in Britain." *Geoscientist* 17, no. 3 (March 2007): The Geological Society. Available online. URL: http://www.geolsoc.org.uk/gsl/cache/offonce/geoscientist/features/pid/882;jsessionid=CC73D9E1F56A559B3729F343E91DBA68. Accessed May 28, 2007. An account of the years Victor Goldschmidt spent working in Britain.

Hughes, Patrick. "Alfred Wegener (1880–1930)." Pangaea. Available online. URL: http://www.pangaea.org/wegener.htm. Accessed October 24, 2007. A brief biography of Wegener.

Israeli Foreign Ministry. "Timna: Valley of the Ancient Copper Mines." Jewish Virtual Library. Available online. URL: http://www.jewishvirtuallibrary.org/jsource/Archaeology/timna.html. Accessed June 11, 2007. A description of how copper was mined and processed from the fifth century B.C.E.

Kennedy, D. J. "St. Albertus Magnus." In *Catholic Encyclopedia,* vol. 1. Christian Classics Ethereal Library. 2005. Available online. URL: http://www.ccel.org/ccel/herbermann/cathen01.html?term=St.%20Albertus%20Magnus. Accessed June 22, 2007. A brief biography of Albertus Magnus.

Lelgemann, Dieter. "On the Ancient Determination of the Meridian Arc Length by Eratosthenes of Kyrene." Athens, Greece: Workshop on the History of Surveying and Measurement, May 22–27, 2004. Available online. URL: http://www.fig.net/pub/athens/papers/wshs1/WSHS1_1_Lelgemann.pdf. Accessed January 24, 2007. An account of the methods Eratosthenes used to measure the circumference of the Earth.

Lendering, Jona. "Hecataeus of Miletus." Livius.org. Available online. URL: http://www.livius.org/he-hg/hecataeus/hecataeus.htm. Accessed January 31, 2007. A brief biography of Hecataeus.

Natland, James H. "James Dwight Dana (1813–1895): Mineralogist, Zoologist, Geologist, Explorer." *GSA Today* 13, no. 2 (February 2003): The Geological Society of America. Available online. URL: http://gsahist.org/gsat/gt03feb20_21.pdf. Accessed September 16, 2008. A brief biography of Dana.

Nave, C. R. "HyperPhysics." Department of Physics and Astronomy, Georgia State University. Available online. URL: http://hyperphysics.phy-astr.gsu.edu/hbase/hph.html. Accessed October 1, 2007. An educational site providing access to information on every branch of physics, including geophysics.

O'Connor, J. J., and E. F. Robertson. "Albertus Magnus." School of Mathematics and Statistics, University of St. Andrews. 2003. Available online. URL: http://www-groups.dcs.st-and.ac.uk/~history/

Biographies/Albertus.html. Accessed June 22, 2007. A brief biography of Albertus Magnus.

———. "Hipparchus of Rhodes." Scotland: School of Mathematics and Statistics, University of St. Andrews. 1999. Available online. URL: http://www-history.mcs.st-andrews.ac.uk/Biographies/Hipparchus.html. Accessed January 31, 2007. A brief biography of Hipparchus.

———. "Jean Richer." School of Mathematics and Statistics, University of St. Andrews. 1996. Available online. URL: http://www-history.mcs.st-andrews.ac.uk/Biographies/Richer.html. Accessed January 29, 2007. A brief biography of Richer.

———. "Regnier Gemma Frisius." School of Mathematics and Statistics, University of St. Andrews. 1996. Available online. URL: http://www.history.mcs.st-andrews.ac.uk/Biographies/Gemma_Frisius.html. Accessed February 6, 2007. A brief biography of Frisius.

———. "Willebrord van Royen Snell." School of Mathematics and Statistics, University of St. Andrews. 1996. Available online. URL: http://www-history.mcs.st-andrews.ac.uk/Biographies/Snell.html. Accessed February 6, 2007. A brief biography of Snell.

———. "Zhang Heng." School of Mathematics and Statistics, University of St. Andrews, 1996. Available online. URL: http://www-history.mcs.st-andrews.ac.uk/Biographies/Zhang_Heng.html. Accessed March 7, 2007. A brief biography of Zhang.

Riebeck, Holli. "Paleoclimatology: Introduction." Earth Observatory, NASA, June 28, 2005. Available online. URL: http://earthobservatory.nasa.gov/Study/Paleoclimatology/. Accessed September 14, 2007. An introduction to the development of paleoclimatology.

Scripps Institution of Oceanography. "Robert Sinclair Dietz Biography." Available online. URL: http://scilib.ucsd.edu/sio/archives/siohstry/dietz-biog.html. Accessed October 25, 2007. A brief biography of Dietz.

Sedlacek, Cheryl. "The Great Glacier Controversy." In *History of Geology*. Emporia State University. Available online. URL: http://www.emporia.edu/earthsci/student/sedlacek1/website.htm. Accessed September 13, 2007. An account of the development of the realization that glaciers had once extended much farther than they do now.

Silkroad Foundation. "Marco Polo and His Travels." Available online. URL: http://www.silk-road.com/artl/marcopolo.shtml. Accessed July 3, 2007. An account of the travels of Marco Polo.

Soylent Communications. "William Smith." NNDB. Available online. URL: http://www.nndb.com/people/264/000106943/. Accessed September 24, 2007. A brief biography of Smith.

Suong Su. "Johann Friedrich Blumenbach 1752–1840." EMuseum@ Minnesota State University, Mankato. Available online. URL: http://www.mnsu.edu/emuseum/information/ biography/abcde/ blumenbach_friedrich.html. Accessed September 19, 2007. A brief biography of Blumenbach.

United Nations Atlas of the Oceans. "Christopher Columbus (1451–1506) Biography." Available online. URL: http://www.oceansatlas. org/unatlas/uses/transportation_telecomm/maritime_trans/nav/ christopher_bio.htm. Accessed April 8, 2008. A brief biography of Columbus.

United States Geological Survey, with U.S. Minerals Management Service. "Methane Gas Hydrates." January 23, 2007. Available online. URL: http://geology.usgs.gov/connections/mms/joint_projects/methane. htm#top. Accessed June 25, 2007. A description of methane hydrates, how they form, where they occur, and their economic and environmental importance.

University of California, Berkeley. "Alfred Wegener (1880–1930)." Available online. URL: http://www.ucmp.berkeley.edu/history/wegener. html. Accessed October 23, 2007. A brief biography of Wegener.

Watson, J. M. "Alfred Lothar Wegener: Moving Continents." U.S. Geological Survey. May 5, 1999. Available online. URL: http://pubs.usgs. gov/gip/dynamic/wegener.html. Accessed October 23, 2007. A brief biography of Wegener, emphasizing his theory of continental drift and linking to explanations of plate tectonics.

INDEX